HOW TO FIND OUT ABOUT PHYSICS

*A Guide to Sources of Information
arranged by the Decimal Classification*

BY

B. YATES, B.Sc.

PERGAMON PRESS

OXFORD · LONDON · EDINBURGH · NEW YORK
PARIS · FRANKFURT

Pergamon Press Ltd., Headington Hill Hall, Oxford
4 & 5 Fitzroy Square, London W.1

Pergamon Press (Scotland) Ltd., 2 & 3 Teviot Place, Edinburgh 1

Pergamon Press Inc., 122 East 55th Street, New York 10022

Pergamon Press GmbH, Kaiserstrasse 75, Frankfurt-am-Main

Set in 10 on 12 pt Times
and Printed in Great Britain
by Billing & Sons Limited, Guildford and London

(2319/65)

Contents

List of Illustrations

Preface

IN ORDER to keep the price of publication of this series low, it is necessary to print a large edition and to aim at several levels of readership. It is hoped that this book will be of sufficient interest at every level to justify purchase, but it cannot be completely satisfactory at any one level. It is only designed to meet some of the needs of students, physicists and the general public, and no claim is made to comprehensiveness; indeed, the deliberate policy has been to give indicative selections only, based on my experience as an information scientist. Nevertheless, I hope that the book will also be helpful to the librarian, although it is not bibliographically comprehensive or annotated in detail. In the main, the material quoted was published prior to 1963.

The appropriate Dewey Decimal Classification class numbers are quoted in order to familiarize the user with the scheme most likely to be found in libraries or reference works. It is, however, recognized that the Dewey Classification is not an ideal arrangement, but it has the great advantage that it is extensively used, and its inconsistencies are well known.

In the Dewey Decimal Classification and its extension the Universal Decimal Classification (UDC) knowledge is divided into nine classes, a tenth class being for general works. Each class is sub-divided decimally. Pure Science 500 has a sub-division Physics 530. Further sub-division leads to numbers for the various fields of physics, e.g. 534 Sound, 539.7 Nuclear Physics. More specific aspects are denoted by yet further sub-division. The Dewey common sub-divisions have also been used to group material, e.g. .02 handbooks, .03 encyclopaedias, dictionaries, .04 essays, reports, .05 periodicals, abstracts, .06 societies, etc.

The American pattern of education often requires "a text of the course". Some questions, therefore, follow each chapter as a possible help to anyone using this publication as a text for teaching sources of information.

Earlier publications concerned with the literature of physics are:

PARKE, N. G., *Guide to the Literature of Mathematics and Physics*, Dover Publications, New York, 2nd ed., 1958.

WHITFORD, R. H., *Physics Literature*, Scarecrow Press, London, 1954.

Parke is a classified guide to the literature addressed to scientists, librarians, engineers and students, and contains a suggested method of searching the literature, whilst Whitford is a guide to the literature at college level. The present volume is more up to date and selective and includes material on careers, societies and non-literature sources of information.

The completion of this volume would not have been possible without the help and co-operation of many people, in particular Dr. G. Chandler, Mr. G. I. Maughan, Mr. A. F. Armstrong, Mrs. S. A. Lowndes and Mrs. C. E. Rigby.

B. Y.

CHAPTER I

Careers in Physics

NATURAL philosophy was the original term for the study of what we now call physics, which thus arose from the study of nature in general. Now physics is more properly defined as the science of material inorganic nature and is concerned with the theory of the general laws of nature. Physics is fundamental to many branches of science, and industry has relied greatly on the part played by physicists in overcoming production problems.

The U.S. National Roster of Scientific and Specialist Personnel defined physics in the *Review of Scientific Instruments*, 15 February 1944, as "the science that deals with those phenomena of inanimate matter involving no change in chemical composition and more specifically it is the science that deals with matter and motion. Recognized areas of specialization within this field are mechanics, heat, sound, light, electricity and magnetism, electronics and ionics, radio, atomic and nuclear physics, properties of materials, theoretical physics and biophysics. Other specialities relate to the application of the fundamentals of science to industrial problems, especially with highly precise and delicate measuring instruments, radio design and manufacture, optical instruments and physical testing of materials."

The range of physics, so far as this work is concerned, is that adopted in the Decimal Classification's 530 class, plus 548 class.

A recent publication by the Ministry of Labour Central Youth Employment Executive, *Careers Guide*, HMSO, London, 5th ed., 1962, states that "physics deals with the properties of matter and energy", "provides the theoretical basis of engineering science" and "ranges from practical experimentation to what is hard to distinguish from mathematics".

There is now virtually no field of industry that does not or could not employ a physicist in some form or another. Progress will be dependent on academic and personal qualities. Each branch of the profession requires different qualities. Before embarking on one particular type of work, the advice should be sought of some-one with experience in that field. Amongst those willing and com-petent to give advice are schoolmasters, lecturers, professors, university appointments boards, local education authorities, the Youth Employment Service, the local or national body concerned with physics and some of the Ministry of Labour Employment Exchanges.

As industry becomes more aware of the benefits of science and technology the need for physicists grows. Up to the end of the nineteenth century there were few physical laboratories in exis-tence. Practically the only opportunities open to graduates were as teachers in universities or in the larger public schools. The idea that physicists could be employed outside the academic sphere was very slow to be accepted. Professor J. A. Crowther, in 1936, stated at a conference of leading employers of scientists and engineers that "the title 'physicist' is still not understood by a large section of the general public". Even at that conference several of the speakers asked for a definition of a physicist!

Guides to Scholarships, Awards, etc.

Having decided to undertake a course of study in physics the student is naturally interested in finding out what financial help is available to him. Financial help for students in the United States is described in the following:

MATTINGLY, R. C., *Financial Aid for College Students: Graduate*, U.S. Government Printing Office, Washington D.C., 1957. (U.S. Office of Education Bulletin, No. 17, 1957.)

WILKINS, F. B., *Financial Aid for College Students: Undergraduate*, U.S. Government Printing Office, Washington D.C., 1957. (U.S. Office of Education Bulletin, No. 18, 1957.)

The National Academy of Sciences Research Council, 2101 Constitution Avenue, Washington, administers the National Research Council Science fellowships, Medical fellowships, Merck fellowships in natural sciences, RCA fellowships in electronics, Fulbright scholarships programme, National Science Foundation fellowships and Lily fellowships in the natural sciences. Further details of these awards can be obtained from the Academy.

Details of awards can also be found from M. A. Firth (ed.), *Handbook of Scientific and Technical Awards in the United States and Canada, 1900–1952*, Special Libraries Association, New York, 1956.

Students, other than U.S. Nationals, who wish to study in the United States can obtain details of the various awards open to them from the U.S. Embassy in their country.

Good outlines of scholarships available in the United Kingdom are described in the following:

MERRIMAN, A. D., *Careers in Science and Technology*, Associated Newspapers, London, 1960. This covers university awards, government administered awards, private trusts and foundations, state scholarships and local education authority awards.

WHEATLEY, D. E., *Industry and Careers*, Iliffe Books, London, 1961. In the chapter entitled "How to Find Out" there is a section on "Financial Assistance for Students" giving details of university scholarships and exhibitions, state scholarships, local education authority awards, overseas scholarships and university awards by other organizations.

A valuable book published by the National Union of Students is the *Grants Yearbook: Local Education Authority Awards to Students*, National Union of Students, London. Local Authorities' awards, post-graduate awards such as DSIR studentships and NATO studentships are described.

The Department of Scientific and Industrial Research makes research grants and information about these is available in *D.S.I.R. Research Grants*, HMSO, London, and in *D.S.I.R. Studentships and Fellowships*, HMSO, London.

Scholarships offered to British students by foreign government. and universities are listed in the yearly publication *Scholarships Abroad*, British Council, 65 Davies Street, London, W.1, whilst an international handbook on scholarships, fellowships and educational exchanges published yearly is *Study Abroad*, United Nations Educational, Scientific and Cultural Organization, Paris.

Published bi-annually is *United Kingdom Postgraduate Awards*, Association of Universities of the British Commonwealth, 36 Gordon Square, London, W.C.1. This lists post-graduate awards for scholarships, fellowships, grants, etc. Australia, Ceylon, East Africa, Ghana, Federation of Rhodesia and Nyasaland, Pakistan, Federation of Malaya, Canada, India, New Zealand and the United Kingdom award Commonwealth fellowships for post-graduate study. Information can be obtained from the Commonwealth Scholarship Commission, Association of Universities of the British Commonwealth, 36 Gordon Square, London, W.C.1.

A handbook detailing opportunities for U.S. nationals is *Handbook of International Study for U.S. Nationals*. It has sections on education abroad; study awards and special programmes for U.S. nationals; summer opportunities abroad for U.S. nationals; organizations in the U.S. providing services to Americans going abroad; U.S. information centres and binational centres abroad; diplomatic missions, government information and tourist offices in the U.S. It is published by the Institute of International Education, New York. A similar publication for other than U.S. nationals is *Handbook for International Study for Foreign Nationals*, Institute of International Education, New York.

A recent publication, which includes details of scholarships for the universities of western Europe, is P. Latham's *Student's Guide to Europe*, MacGibbon & Kee, London, 1962. As well as describing methods of application for entry, the book summarizes the educational systems of each country.

Guides to Careers

Guides to careers exist in three forms: those covering all types of careers, those covering physics in particular and those covering specifically one of the various branches of physics.

Amongst the guides covering all types of careers are:

BATTEY, E. W., *Scientific and Technical Education and Careers*, Herbert, London, 1959.

CHAFFE, G. H. and EDMONDS, P. J., *Careers Encyclopaedia*, Cleaver-Hume, London, 2nd ed., 1958.

GREENLEAF, W. J., *Occupations and Careers*, McGraw-Hill, New York, 1955.

MERRIMAN, A. D., *Careers in Science and Technology*, Associated Newspapers, London, 1960.

POLLACK, P., *Careers Opportunities in Science*, Dutton, New York, 1954.

PRINGLE, P., *Careers Handbook*, Ward Lock, London, 1959.

Opportunities for Girls and Women in Science and Technology, British Federation of University Women, London.

Scientific and Technical Education and Careers, Herbert, London. 1959.

Scientific Careers for Women, HMSO, London.

Scientist, HMSO, London. (Choice of Careers, Series No. 100.)

The National Union of Teachers Annual Guide to Careers for Young People, National Union of Teachers, London.

The Times Book of Careers in Industry, Blackie, London, 1959.

The Central Youth Employment Agency is responsible for *Careers Guide: Opportunities in the Professions, Industry and Commerce*, HMSO, London, 5th ed., 1962. This gives details of over 120 careers, of which physics is one, giving a summary of the type of work, qualities and educational qualifications required, training, opportunities and prospects, and sources of further information.

The sort of information obtainable from guides of this type is well illustrated by *Industry and Careers*, Iliffe, London, 1961. It contains sections on planning a career; the way in; opportunities for women in industry; agriculture; horticulture; forestry and related services; mining and quarrying; food, drink and tobacco; chemicals and allied industries; metal manufacture; engineering, shipbuilding and ship repairing; textiles; leather; clothing and footwear; bricks, pottery, glass, cement, etc.; furniture; paper and printing; other manufacturing industries; construction; gas and electricity; transport and communications; specialist services; further education; how to find out.

For industrial posts in Great Britain a good guide is the *Directory of Opportunities for Graduates*, Cornmarket Press, London. This is published every year and in it several hundreds of the major concerns give details of their requirements indicating the type of work to be done, conditions of service and the type of person they are looking for. Other generally valuable guides in this series are: *Directory of Opportunities for School Leavers* and *Directory of Opportunities for Qualified Men*. Companion volumes exist for Canada—*Canada Careers Directory* and *School Careers Directory;* and France—*Choisissez votre Carrière*.

Published each mid-term is *Careers in Government, Industry and Commerce*, Careers Press, Croydon. Under occupational headings, such as "physicist", are listed the names of various companies requiring staff with an indication of the educational standards required. This publication is specially designed for the student.

Guides devoted to physics include N. Clarke's *Physics as a Career*, Chapman & Hall, London, 2nd rev. ed., 1959. This is the best publication covering the United Kingdom and has chapters on physics and its applications; professional education and qualifications; industrial posts; university posts; government service; the United Kingdom Atomic Energy Authority; technical college posts; school teaching; Research Association posts; hospital physicists; other posts; salaries; on finding a job; institutions and societies.

Covering the United States, and still of use despite its age, is A. W. Smith's *Careers in Physics*, Longs College, Columbus, Ohio, 1951. As an indication of its content the following are its chapter headings: physics as a profession; educational preparation and goals; supply and demand; careers in education; industrial physicists; power and progress; illumination and vision; communication, transportation and mobility; materials and prospects for better living; photography and its uses; instrumentation; careers in Federal research agencies; careers in military research agencies; research institutes and foundations; classical physics; modern physics; atmospheric physics and meteorology; food and health physics; borderland fields for specialization; industrial research laboratories; bibliography.

Other books concerning physics careers in the United States include:

Career as a Physicist, Institute for Research, Chicago, 1946.
Manpower Resources in Physics, U.S. Government Printing Office, Washington D.C., 1952. (Scientific Manpower Series No. 3.) This describes the training, specialization, age, distribution and employment of 6600 physicists.
Physics as a Career, American Institute of Physics, New York, 1952.

Interesting articles often appear in the physics periodicals which throw light on some aspects of physics as a career. M. W. White's article "American Physicists in the Current Quarter Century", *Phys. To-day*, **9**, 31 (January 1956), surveyed the educational training, occupational specialities, employers, functions performed, work classification and salaries of physicists in the United States; C. G. Suit's article "A Future for Physicists in Industry", *Phys. To-day*, **9**, 28 (January 1956), examined the status of the industrial physicist and assessed the opportunities for the future.

The leading newspapers and periodicals such as *Nature, Nucleonics, New Scientist, Institute of Physics Bulletin, Physics To-day, Journal of Applied Physics*, etc., carry advertisements for vacant physics positions which give a general indication of the work involved and the qualifications called for.

Research

The biggest single employer of physicists doing research work will probably always be the Government. The Armed Services, the Atomic Energy Authority, the Central Electricity Generating Board, the Post Office, etc., all undertake research and employ physicists. Useful background information can be obtained from *Government Scientific Organization in the Civilian Field*, HMSO, London, 1961. The addresses of all the governmental laboratories can be found in both *Industrial Research in Britain*, Harrap, London, 4th ed., 1962, and *Scientific and Learned Societies of Great Britain*, Allen & Unwin, London, 1962.

For the United States it is interesting to read *Federal Organization for Scientific Activities, 1962*, U.S. Government Printing Office, Washington D.C., 1963. This is a report of the organization and programme content for science of the forty Federal agencies involved in scientific activities. The report on each government unit covers research and development, scientific and technical information, research data collection and scientific testing and standardization.

Of some interest to those contemplating working for the U.S. Government is the survey compiled by the Maxwell Graduate School of Citizenship and Public Affairs for the Office of National Research—*Attitudes of Scientists and Engineers about their Governmental Employment*, Syracuse University, Syracuse, New York, 1950.

A guide to research laboratories in the United Kingdom is *Industrial Research in Britain*, Harrap, London, 4th ed., 1962. This details the research work of government laboratories, universities, technical colleges, industry and other non-governmental laboratories. There is a subject index. The work also contains an article on "Careers in Professions associated with Industrial Research".

For the United States a useful guide is *Industrial Research Laboratories of the United States*, National Academy of Sciences, Washington D.C., 11th ed., 1960. This lists non-governmental laboratories concerned with fundamental and applied research including development of products and processes. Information on 5420 laboratories is given and an individual name index is included.

Atomic Energy

One field which requires a considerable number of physicists is atomic energy. It is probably true to say that in the past few years this has been the "glamour" industry for physicists.

In Great Britain, the United Kingdom Atomic Energy Authority, Charles II Street, London, has issued several booklets of interest, amongst them being:

Careers in Reactor Technology.
Careers for Youth in Scientific Research.
Scientific Appointments in the Research Group of the Atomic Energy Authority.

T. B. Le Cren has an article on "The United Kingdom Atomic Energy Authority" in *Scientific and Technical Education and Careers*, Herbert, London, 1959. Interesting background reading is *Britain's Atomic Factories*, HMSO, London, 1954; *Harwell, the British Atomic Energy Research Establishment, 1946–1951*, HMSO, London, 1952; JAY, K. E. B., *Atomic Energy Research at Harwell*, Butterworths, London, 1955, and *The Nuclear Energy Industry of the United Kingdom*, UKAEA, London, 2nd ed., 1961.

A good indication of industrial firms who are working in this field is given in the various Buyers' Guides. For Great Britain these are:

Nuclear Engineering Buyer's Guide. This is included in the March 1962 issue of the journal *Nuclear Engineering.*
Nuclear Power Industry Guide and Digest, Rowse Muir, London, 1961.

For the United States the best guide is:

Nucleonics Buyers' Guide which forms part of the November issue of the journal *Nucleonics* each year.

One must also mention in connection with nuclear research in Great Britain the National Institute for Research in Nuclear Science, Rutherford High Energy Laboratory, Harwell, Didcot, Berkshire. The laboratory was opened in 1959. The Government recognized that the great cost of advanced equipment in the nuclear field put research beyond the means of an individual university or institution. As a result it established this Institute to provide facilities for common use. The governing body includes representatives of universities, the Royal Society, the Atomic Energy Authority and the Department of Scientific and Industrial Research. A 7 GeV proton synchroton is being constructed and a 50 MeV proton linear accelerator is already operating. A second laboratory is soon to be in existence at Daresbury, Cheshire.

Almost every country has a national body concerned with atomic energy, and several private firms as well. Then there are the international bodies such as the International Atomic Energy Agency (IAEA), Kaertnerring, Vienna; CENTO Institute of Nuclear Science, Tehran; European Atomic Energy Community (EURATOM), 51 Rue Belliard, Brussels; Organization Européenne pour la Recherche Nucléaire (CERN), Meyria, Geneva; OECD European Nuclear Energy Agency (ENEA), 38 Boulevard Suchet, Paris 16; Foratom Atomique Européen (FORATOM), 4 Rue de Tehran, Paris. Any of these bodies should be pleased to provide details of openings for physicists and the qualifications required.

Details of the organizations working in atomic energy can be found in *World Nuclear Directory*, Harrap, London, 2nd ed., 1963. This lists over 2000 organizations such as agencies, authorities, boards, commissions, committees, societies, government departments, divisions of the armed services, insurance and power groups, universities and industrial firms. As well as the full title and address of the organization the names of senior personnel and members of appropriate committees are given.

International Bodies

International bodies which employ scientists include the United Nations Educational, Scientific and Cultural Organization (UNESCO), the World Health Organization (WHO) and the World Meteorological Organization (WMO). Others which have already been mentioned under atomic energy are IAEA, EURATOM, CERN, etc.

UNESCO came into being on 4 November 1946. Its purpose is to contribute to peace and security by promoting collaboration among the nations through education, science and culture.

In the science field, UNESCO seeks to promote international scientific co-operation by instituting meetings between scientists and aiding the works of international scientific organizations. It encourages scientific research designed to improve the living conditions of mankind. Science co-operation offices have been set up in Montevideo, Cairo, New Delhi and Jakarta. In its mass communication work it disseminates information, carries out research and provides advice. National commissions act as liaison groups between UNESCO and the educational, scientific and cultural life of their own countries.

For background information on UNESCO one should read *Science Liaison*, UNESCO, Paris, 1954, which is the story of UNESCO's Science Liaison Offices, and LAVES, W. H. C. and THOMSON, C. M., *UNESCO—Purpose, Progress and Prospects*, Dobson, London, 1958.

Details of the various international scientific organizations can be obtained from guides such as:

Directory of International Scientific Organizations, UNESCO, Paris, 1950.

Guide to International Organizations, British Central Office of Information, London, 1953–.

Yearbook of International Organizations, Union of International Organizations, Brussels.

Questions

1. What opportunities exist in industry for the physicist?
2. What opportunities exist in the Government Service for the physicist?
3. What opportunities exist in international bodies for the physicist?

CHAPTER 2

Books: General, Textbooks, Monographs, Series and Surveys of Progress

Dewey 530 Class, etc.

How to Evaluate and Trace Physics Books

One has to qualify any recommendation of a book by saying that this is the best book for beginners, this is the best book for graduates, and so on. Even then what suits one person may not suit another. The worth of any publication very much depends on the user, but a general assessment can be made if the following points are borne in mind:

(a) The author—is he a well-known name in the field, has he written other books?

(b) The content of the book—this can be gauged from the chapter headings, the subject index, and by scanning the pages and index.

(c) The presentation and treatment—is the layout of each page pleasing, are graphs and tables easily readable, does the writing style appeal?

(d) Date of publication and edition—if there have been several editions of the book, this indicates that a large number of people have found it valuable.

Naturally, the later the date of publication of a book the more up to date it should be. Other things being equal, this would decide the choice between particular publications. Wide though the selection of books may be in any library, few libraries can buy all publications. After looking through the catalogue at a particular

library, it may be necessary to consult other catalogues to check
if more publications exist covering your interest.

Catalogues of the various national libraries will prove of value,
especially *A Catalog of Books Represented by the Library of Congress Printed Cards*, which was published in 167 volumes and has
been kept up to date by supplements, and the catalogue of the
British Museum which is in course of publication. As well as
published catalogues, lists of published books are available for
consultation. *Cumulative Book Index* is published monthly and
regularly cumulated by H. W. Wilson, New York, and is a world
list of books in the English language with entries under the
author's name, subject and title. The *British National Bibliography*
is published weekly with regular cumulations and is more comprehensive with British publications than the *Cumulative Book Index*.
Arrangement is by the Dewey Decimal Classification. R. L.
Collinson's *Bibliographical Services Throughout the World*,
UNESCO, Paris, 1961, can be used to trace other national bibliographies.

Special catalogues of scientific books include:

Annual Physics Book List, American Institute of Physics, New York.

British Scientific and Technical Books, 1935–1952 and 1953–1957,
Aslib, London, 1956–60. *Aslib Booklist* keeps the basic volumes
up to date.

*Catalogue of Lewis's Medical, Scientific and Technical Lending
Library*, Lewis, London, 1957. Supplements cover 1957–59,
1960.

*Scientific, Medical and Technical Books Published in the United
States of America*, National Research Council, Washington
D.C. The basic volume, edited by R. R. Hawkins, is kept up to
date by supplements.

Selected List of Standard British Scientific and Technical Books,
Aslib, London, 6th ed., 1962.

Technical Book Review Index, U.S. Special Libraries Association,
Chicago. Issued ten times per year.

Useful for keeping up to date with the latest publications are the book reviews in the scientific periodicals such as *Physics Today*, *New Scientist* and *Nature*.

General Textbooks on Physics [Dewey 530 Class]

General textbooks introduce physics to the uninitiated and their contents range over the whole spectrum of physics, or at least over several of its branches.

What is Science? Gollancz, London, 1956, is an explanation of the various scientific fields by eminent scientists and could well serve as an introductory text. Another, equally suitable, is one of the many publications by E. N. da C. Andrade, *An Approach to Modern Physics*, Bell, London, 3rd ed., 1962; or W. Railston's *Teach Yourself Physics*, English Universities Press, London, 1960; or The Physical Science Study Committee's improved beginning course—*Physics*, Heath, Boston, 1960. Other suitable introductory books include:

BEISER, A., *The Mainstream of Physics*, Addison-Wesley, Reading, Massachusetts, 1962.

DULL, C. E., *et al.*, *Modern Physics*, Holt, New York, 1960.

FREEMAN, I. M., *Modern Introductory Physics*, McGraw-Hill, New York, 2nd ed., 1957.

GAMOW, G. and CLEVELAND, J. M., *Physics Foundations and Frontiers*, Prentice-Hall, London, 1960.

KIMBALL, A. L. and WATERMAN, A. T., *College Textbook of Physics*, Holt, New York, 6th ed., 1954.

SEARS, F. W. and ZEMANSKY, M. W., *College Physics*, Addison-Wesley, Reading, Massachusetts, 3rd ed., 1960.

SHORTLEY, G. and WILLIAMS, D., *Principle of College Physics*, Prentice-Hall, Englewood Cliffs, New Jersey, 1959.

VAN NAME, F. W., *Modern Physics*, Prentice-Hall, Englewood Cliffs, New Jersey, 2nd ed., 1962.

WATTS, D. S., *Physics is Easy*, Princes Press, London, 1959.

Those studying for the General Certificate of Education at ordinary level have many textbooks to choose from, such as

C. W. Kearsey's *A School Physics*, Longmans, London, 1956. For
the student taking the advanced level examinations any of the
following may be used:

BROWN, R. C., *A Textbook of Physics*, Longmans, London, 1961.

NEWMAN, F. H. and SEARLE, V. H. L., *The General Properties of
Matter*, Longmans, London, 5th ed., 1959.

NOAKES, G. R., *New Intermediate Physics*, Macmillan, London,
1957.

SMITH, C. J., *Intermediate Physics*, Arnold, London, 4th ed., 1957.

Taking the subject to scholarship standard is D. H. Fender's
General Physics and Sound, English Universities Press, London,
1957.

University students, past and present, all testify to the worth of
S. G. Starling's and A. J. Woodall's *Physics*, Longmans, London,
2nd ed., 1957, whilst another running into several editions and
based largely on lectures given to honours students is H. A.
Wilson's *Modern Physics*, Blackie, London, 4th ed., 1959. R.
Kronig's *Textbook of Physics*, Pergamon, Oxford, 2nd ed., 1960,
surveys the whole field.

Series of Physics Monographs

Several publishers publish important series of value to students
of physics, such as:

Cambridge University Press. Cambridge Monographs on Physics.
Heinemann: Science Study Series, which is an outcome of the work
of the Physical Science Study Committee. Titles include:
Gravity.
Magnets.
Near Zero.
The Physics of Television.
Holt, Rinehart and Winston's series explaining science in a simple
manner, which includes:
Modern Physical Science.
Modern Physics.

McGraw-Hill International Series of Pure and Applied Physics.

McGraw-Hill. National Nuclear Energy Series. This series has the authorization of the U.S. Atomic Energy Commission.

Methuen's Monographs on Physical Subjects, which are more specialized.

Pergamon International Series of Monographs on Physics. The Commonwealth and International Library of Science, Technology and Engineering, has a number of specialized divisions of interest to physicists.

Prentice-Hall Science Series, which includes:

MARCUS, A., *Physics for Modern Times.*

PELLA, M. O. and WOOD, A. G., *Physical Science for Progress.*

Van Nostrand's series of introductory texts, which includes:

Physics as a Basic Science.

Physics as an Exact Science.

Physical Science as a Basic Course.

Surveys of Progress in Physics

Several reviews of progress in various fields are published, usually annually, and bridge the gap between the textbook and the journal, for example:

Advances in Chemical Physics, Interscience, New York. An annual publication, Volume III being published in 1961 and covered mechanism of organic electrode reactions; non-linear problems in thermodynamics of irreversible processes; propagation of flames and detonations; large tunnelling corrections in chemical reaction rates; recent aspects of diamagnetism; power electrodes and their applications; variational principles in thermodynamic and statistical mechanics of irreversible processes; electron diffraction in gases and molecular structure.

Advances in Science. Monographic reviews which are issued by the U.S.S.R. All Union Institute for Scientific and Technical Information.

HASLETT, A. W. and ST. JOHN, J., *Science Survey*, Vista Books, London. An annual survey which records recent scientific achievements and advances.

Reports on Progress in Physics, Institute of Physics and the Physical Society, London. Experts report on recent advances. Each article usually contains a good bibliography. Issued annually since 1934 except for the period 1942–49. Author and subject index have been issued for the period 1934–52.

Science in Progress, Yale University Press, New Haven, Connecticut. A series of biennial volumes which presents the reports of eminent scientists on the latest developments in their fields.

SHAMOS, M. H. and MURPHY, G. M. (eds.), *Recent Advances in Science, Physics and Mathematics*, Interscience, New York, 1956. This treatment presupposes some scientific training and the coverage includes applied mathematics, operations research, atomic structure, microwave spectroscopy, nuclear structure and transmutations, elementary particles, electronuclear machines, neutron physics, transistor physics, ferromagnetism, cryogenics, physics for the engineer.

Solid State Physics—Advances in Research and Applicatio Academic Press, New York, Issued annually since 1955.

For books on specialized branches of physics see the appropriate chapter.

Questions

1. How can one keep in touch with the latest physics textbooks?
2. Write short notes on one of the most interesting developments in physics during the last five years.

CHAPTER 3

Handbooks, Tables, Dictionaries, Encyclopaedias, Biographical and Historical Reference Books

Dewey 530.2, 530.3 Classes, etc.

Handbooks and Tables

There are a number of outstanding handbooks of physics, including:

CONDON, E. U. and ODDISHAW, H. (eds.), *Handbook of Physics*, McGraw-Hill, New York, 1958. A one volume synthesis of the principle parts of the science. Each section is written by an expert.

FLÜGGE, S. (ed.), *Handbuch der Physik*, Springer, Berlin, 1955–62. This is a new edition of 54 volumes. Each section is written by an expert in the field and in most cases contains a bibliography. The series covers mathematical aids—two volumes; principles of theoretical physics—three volumes; mechanical and thermal properties of matter—ten volumes; electrical and magnetic properties of matter—eight volumes; optics—six volumes; Roentgen rays and particle rays—five volumes; atomic and molecular physics—three volumes; nuclear physics—eight volumes; cosmic rays—one volume; geophysics—three volumes; astrophysics—five volumes.

GRAY, D. E. (ed.), *American Institute of Physics Handbook*, McGraw-Hill, New York, 2nd ed., 1963. Has sections on aids to computation, mechanics, acoustics, heat, electricity and magnetism, optics, atomic and molecular physics, nuclear physics, solid state physics.

WIEN, W. and HARMS, F., *Handbuch der experimental Physik*, Akademische Verlagsgesellschaft, Leipzig, 1926–35. 28 volumes.

Often specific facts are required, such as the thermal conductivity of a particular material, and whilst information can be obtained from encyclopaedias, textbooks, periodicals, etc., it is much easier to locate it in compilations of tables. The effectiveness of a particular compilation depends not only on its date of publication but also on its layout and indexes. The enquirer would do well to familiarize himself with these.

Perhaps the best known of all tables is *International Critical Tables*, McGraw-Hill, New York, 1926–33. Issued in seven volumes with an index, this contains numerical data covering physics, chemistry and technology. "Critical" in this connection means that the co-operating expert gave the "best" value he could derive from all the information available, together, where possible, with an indication of its reliability. A more recent publication containing critically evaluated numerical property values is *Consolidated Index of Selected Property Values: Physical Chemistry and Thermodynamics*, National Academy of Science—National Research Council, Washington D.C., 1962. (Publication 976.)

HODGMAN, C. D. (ed.), *Handbook of Chemistry and Physics*, Chemical Rubber, Cleveland, Ohio, is revised yearly and includes mathematical tables; numerical tables; physical constants of elements, inorganic compounds, organic compounds, alloys and plastics; thermodynamic constants of elements, oxides, hydrocarbons; thermal expansion; vapour pressure; heat conductivity; acoustics; velocity of sound and sound absorption; electrical characteristics; units and conversion factors; and miscellaneous basic physical data. This handbook will probably be found in every scientific laboratory and is thus of proved value, as is *Tables of Physical and Chemical Constants*, Longmans, London, 12th ed., 1959, which were compiled by G. W. C. Kaye and T. H. Laby.

The major German publication is by M. H. Landolt and R. Börnstein—*Zahlenwerte und Funktionen aus Physik, Chemie, Astronomie, Geophysik und Technik*, Springer, Berlin, 6th ed.,

1950. It is issued in several volumes covering atomic and molecular physics; characteristics of matter in bulk; astronomy and geophysics; technology. The introductory sections are so comprehensive that they could well be used as textbooks of physics. Exhaustive literature references are also included so that the original source of the data can be consulted.

Amongst other tables of value are:

CHILDS, W. H. J., *Physical Constants Selected for Students*, Methuen, London, 8th ed., 1958.

Documenta Geigy Scientific Tables, Geigy, Basle, Switzerland, 5th ed., 1959. This aims to provide the physician and research worker with basic scientific data in the fields of medicine, biology, chemistry, physics and mathematics in a concise form. In the section of physical tables are units of measurements; time, temperature, mechanical, electrical, magnetic, X-rays, gamma-rays, corpuscular radiation, photometric, acoustic, chemical and physical constants.

FRIEND, J. NEWTON, *Science Data*, Griffin, London, 4th ed., 1960.

LANGE, N. A. (ed.), *Handbook of Chemistry*, Handbook Publishers, Sandonsky, Ohio, 8th ed., 1952. A reference volume for all requiring ready access to chemical and physical data.

Smithsonian Physical Tables, Smithsonian Institute, Washington D.C., 9th ed., 1954.

Tables Annuelles de Constantes et Données Numériques de Chimie, de Physique et de Technologie, Gauthier-Villars, Paris, 1912–37.

Tables de Constantes et Données Numériques de Chimie, de Physique, de Biologie et de Technologie, Hermann, Paris, 1947–52.

Handbook of Thermophysical Properties of Solid Materials, Macmillan, New York, 1961.

Much useful data can also be found in the *American Institute of Physics Handbook*, McGraw-Hill, New York, 2nd ed., 1963, edited by D. E. Gray.

Scientific Dictionaries

The general scientific dictionaries include:

BEADNELL, C. M., *Dictionary of Scientific Terms*, Watts, London, 2nd ed., 1942.

FLOOD, W. E., *Scientific Words, their Structure and Meaning*, Oldbourne, London, 1960.

FLOOD, W. E. and WEST, M., *An Elementary Scientific and Technical Dictionary*, Longmans, London, 3rd ed., 1962.

ZIMMERMAN, O. T. and LAVINE, I., *Scientific and Technical Abbreviations, Signs and Symbols*, Industrial Research Service, Dover, New Hampshire, 2nd ed., 1949.

Chambers's Technical Dictionary, Chambers, London, 1957, is a dictionary which can be seen in scientific establishments everywhere. It aims to give definitions of terms of importance in pure and applied science, in all branches of engineering and construction and in the larger manufacturing industries and skilled trades. It was written by specialists partly for other specialists but more particularly for the technically minded man in the street, and for students and interested workers of all kinds. Very useful also, and not least because of its extensive cross-referencing, is a dictionary by D. W. G. Ballentyne and L. E. Q. Walker, *A Dictionary of Named Effects and Laws in Chemistry, Physics and Mathematics*, Chapman & Hall, London, 2nd ed., 1961. Over 1000 entries, explaining phenomena which are referred to by the name of their discoverer, are included.

Physics Dictionaries

A most useful series of glossaries has been issued by the American Institute of Physics covering terms used in solid state physics, acoustics, physics and computers, nuclear physics, radio astronomy, plasma physics, high energy physics, optics and spectroscopy. The latest and most comprehensive dictionary, indeed as its title suggests, being very close to an encyclopaedia, is edited by

J. Thewlis, *Encyclopaedic Dictionary of Physics*, Pergamon, London, 1961–62, seven volumes. This covers physics proper and to a greater or lesser extent such subjects as mathematics, astronomy, aerodynamics, hydraulics, geophysics, meteorology, physical metallurgy, radiation chemistry, physical chemistry, structural chemistry, crystallography, medical physics, biophysics and photography. An invaluable feature of this publication is that many entries carry references to further reading, thus allowing the enquirer to broaden his knowledge of a particular subject.

Covering laws, relationships, equations, basic principles and concepts as well as the most widely used instruments, apparatus and components is *International Dictionary of Physics and Electronics*, Van Nostrand, Princeton, New Jersey, 2nd ed., 1961. Units and systems of units are treated at length. Background discursive statements and entries as well as formal ones are given. There is a multilingual index in French, German, Russian and Spanish. The introduction reviews developments in physics since 1800.

Earlier dictionaries which may still be useful to find how ideas about particular concepts changed with time include:

GLAZEBROOK, SIR R., *Dictionary of Applied Physics*, Macmillan, London, 1922–23.

GRAY, H. J., *Dictionary of Physics*, Longmans, London, 1958. This publication is extensively cross-referenced and also quotes other references.

WESTPHAL, W. H., *Physikalisches Wörterbuch*, Springer, Berlin, 1952, has appendices giving a history of physics, physicists' life dates and miscellaneous physical tables.

Information on physical laws and their effects can be obtained from a book by C. F. Hix and R. P. Alley, *Physical Laws and Effects*, Wiley, New York, 1958. The book comprises three different cross-reference systems: (1) description of laws and effects with an indication of the expected magnitude and references that will be useful in gaining additional information, (2) cross-reference by fields of science, (3) cross-reference by physical quantities.

B

This section lists not only the law or effect pertaining to the physical quantity but also the other quantities covered by the same law.

Dictionaries which cover the specialized fields of physics such as *Glossary of Terms Used in Nuclear Science—British Standard 3455*, British Standards Institute, London, 1962, will be found in the appropriate chapters. For formulae used in physics one can consult:

MENZEL, D. M. (ed.), *Formulas of Physics*, Prentice-Hall, Englewood Cliffs, New Jersey, 1955.

THOMAS, T. S. E., *Physical Formulae*, Wiley, New York, 1953.

General Dictionaries

Dictionaries, both general and special, should not be overlooked as most useful sources of physics information. They give condensed definitions of physical concepts and can be very helpful as initial steps in any search for information. Using them may help orientate a problem in its particular place in the whole aspect of physics and thus allow a search for information to be rightly directed. When searching for information, two or three dictionaries should be used as, particularly in the cheaper volumes, coverage is sometimes selective. The general dictionaries which will prove useful include:

Oxford English Dictionary.
Webster's New International Dictionary.

Under the auspices of UNESCO a bibliographical guide to monolingual dictionaries has been published: WÜNSTER, E., *Bibliography of Monolingual, Scientific and Technical Glossaries: Volume 1. National Standards*, UNESCO, Paris, 1955. This most useful guide, covering as it does world publications listing graphical symbols, technical terms and standards, will give a lead to the most useful dictionaries.

Foreign Language Dictionaries

More and more physics literature is being published in languages other than English and it is essential if one intends to keep up to date that an attempt should be made to read an article in its original language. Most libraries hold selections of general scientific and technical dictionaries in the more popular languages, examples of which are:

DE VRIES, L., *German–English Science Dictionary*, McGraw-Hill, New York, 1960.

EMEY, M. A. and S. A., *Scientific Russian Guide*, McGraw-Hill, New York, 1961.

GATTO, S., *A Dictionary of Scientific and Technical Terms, Italian–English, English–Italian*, Casa Editrice Ceschina, Milan, 1960.

GUEDECKE, W., *Technische Abkürzungen, Deutsch–English–Französisch*, Brandstetter, Wiesbaden, 1961.

Specialized works can be identified using compilations such as the UNESCO *Bibliography of Interlingual Scientific and Technical Dictionaries*, UNESCO, Paris, 4th ed., 1961. Entries are arranged according to the Universal Decimal Classification and indexes of authors and subjects are in English, French and Spanish. The Library of Congress has also published a useful compilation: *Foreign Language–English Dictionaries, Volume 1: Special Subject Dictionaries, with Emphasis on Science and Technology*, Library of Congress, Washington D.C., 1955. Entries are arranged alphabetically by subject.

Examples of the specialized dictionary are:

KING, G. G., *Dictionnaire Français–Anglais: Électronique Physique Nucléaire*, Dunod, Paris, 1961.

SKIBICKI, W., *Glossary of Physics*, PWN, Warsaw, 1961. Glossary of 7617 Polish terms used in physics with their English, French, German and Russian equivalents.

Encyclopaedias

An encyclopaedia is extremely useful as a source of preliminary knowledge. It should be remembered that an article is necessarily out of date as soon as it is written and thus one should not go to encyclopaedias for very current material. Each article is complete in itself giving a summary of a particular topic. It usually ends with a short bibliography to aid further study. Each article usually consists of a definition, the relation of the topic to other parts of knowledge, the history of the subject, a review of the topic and a bibliography. If the writers of articles in the best encyclopaedias are specialists, their contributions can usually be taken as authoritative. The information in an encyclopaedia is sometimes an end in itself but often it serves to point the way to avenues of solution of particular enquiries. C. M. Winchell usefully evaluates encyclopaedias in her book *Guide to Reference Books*, American Library Association, New York, 7th ed., 1951.

The most famous general encyclopaedia is *Encyclopaedia Britannica*, Encyclopaedia Britannica, Chicago. The 1959 edition has 24 volumes containing 43,000 articles, including many on scientific subjects, each written by experts in the field, with a detailed subject index. Annual supplements reviewing recent advances are issued. Other encyclopaedias include:

Encyclopaedia Americana, Americana Corporation, New York. The 1958 edition has 30 volumes containing 59,000 articles written by experts.

Chambers's Encyclopaedia, Newnes, London, 1959.

Collier's Encyclopaedia, Collier, New York, 2nd ed., 1962.

The more specialized encyclopaedias include *McGraw-Hill Encyclopedia of Science and Technology*, McGraw-Hill, New York, 1960. The basic fifteen volumes of this work are supplemented by yearbooks. The purpose of this work is to provide the widest range of articles useful to a person with some technical training who wishes to obtain information outside his particular field. Biographical and historical articles are not included and the philosophical basis of many subjects is not discussed. *Van Nostrand's Scientific Encyclopedia*, Van Nostrand, Princeton, New Jersey, 3rd ed., 1958, defines about 15,000 terms.

Biographical Dictionaries

The most comprehensive publication for scientists of many nations, giving short collective biographies and including details of the publications of the biographee, is: POGGENDORFF, J. D., *Biographischliterarisches Handwörterbuch der exakten Naturwissenschaften*, Barth, Leipzig, 1863–1904, 1926–40. This covers the period from the earliest times to 1931. Publication recommenced in 1956 and volumes covering the period from 1931 are now being issued. Listing is alphabetical and the fields of mathematics, astronomy, physics, geophysics, chemistry and crystallography are covered.

This is, of course, not the only useful reference tool, amongst the others being:

HEATHCOTE, N. H. DE V., *Nobel Winners in Physics 1901–1950*, Schuman, New York, 1953.

HOWARD, A. V., *Chambers's Dictionary of Scientists*, Chambers, London, 1955.

IRELAND, N. O., *Index to Scientists of the World from Ancient to Modern Times: biographies and portraits*, Saxon, Boston, 1962.

A list of biographies published up to 1948 has been compiled by T. J. Higgins, *Biographies of Engineers and Scientists*, Illinois Institute of Technology, Chicago, 1949, and this can be supplemented by using *Biography Index* which is published quarterly with periodic cumulations by Wilson, New York, and is an index of material which has appeared in books and magazines. It should not be overlooked that *Encyclopaedia Britannica* is a good source of biographical details on deceased scientists and it gives references to the main publications of the biographee. Literally hundreds of biographies on individual scientists have been issued and these should be easily found in a library catalogue or from one of the many listings of book publications such as *British National Bibliography*, British National Bibliography, London.

Guides to those currently engaged in work in the field of physics include *American Men of Science*, Cattell Press, Tempe, Arizona, 10th ed., 1960, edited by J. Cattell. About 120,000 names from the physical, biological, social and behavioural sciences are listed. *Directory of British Scientists*, Benn, London, 1963, is the most recent publication specifically on British scientists and carries over 30,000 names. Covering the Soviet Union is *Who's Who in Soviet Sciences and Technology*, Telberg, New York, 1960.

Some guides cover specific fields such as *Who's Who in Atoms*, Vallancy Press, London, 3rd ed., 1962, which carries over 16,300 names of workers in all parts of the world. Covering the same field but concentrating on the U.S.S.R. is *Who's Who in Soviet Nuclear Science*, Lawrence Radiation Laboratory, Berkeley, California, 1960. The electrical sciences are covered by *Electrical Who's Who 1962–63*, Iliffe, London, 1962, and brief biographies of leading members of the professional and industrial branch of the science are given. The most eminent scientists will, of course, appear in the Who's Who type of publication for their country.

Historical Guides

Reference material dealing with the history of physics includes bibliographies such as *A List of Books on the History of Science*, John Crerar Library, Chicago, 1911. Supplements 1916, 1942–46, F. Russo's *Histoire des Sciences et des Techniques: Bibliographie*, Herman, Paris, 1954, and that appearing in one of George Sarton's books, *A Guide to the History of Science*, Chronica Botanica, Waltham, Massachusetts, 1952. Sections are devoted to international congresses, journals and serials concerning the history and philosophy of science, treatises and handbooks on the history of science, history of special sciences, science and society, scientific societies, institutes, museums and libraries. Sarton is a very prolific writer on the history of science and all his publications can be thoroughly recommended. Covering specifically electricity and magnetism is *Bibliographical History of Electricity and Magnetism*, Griffin, London, 1922, by P. F. Mottelay. This covers the period from 2637 BC to AD 1821.

Chronologies listing the scientific discoveries and inventions of the years include one by L. Darmstaedter, *Handbuch zur Geschichte der Naturwissenschaften und der Technik in Chronologischer Darstellung*, Springer, Berlin, 2nd ed., 1908, which lists discoveries from 3500 BC to AD 1908. A later book by the same author *Entwicklungsgeschichte der modernen Physik*, Springer, Berlin, 1923, carries the tables of scientific discoveries up to 1923.

The many texts covering the history of physics or of specific aspects of physics such as those by Max von Laue, *History of Physics*, Academic Press, New York, 1950, and E. Mach, *The Science of Mechanics: a Critical and Historical Account of its Development*, Open Court, La Salle, Illinois, 1942, can best be found in the same manner as is suggested for biographical material.

For reference books on specialized aspects of physics see the appropriate chapter.

Questions

1. What sort of information can be found from: (*a*) encyclopaedias, (*b*) dictionaries, (*c*) handbooks, (*d*) tables?
2. Compare the information in the *Encyclopaedia of Science and Technology* and in the *Encyclopaedic Dictionary of Physics* on any subject in physics.
3. Use the *Handbook of Chemistry and Physics* to find the surface tension of ethylbenzene.
4. Use the *International Critical Tables* to find the viscosity of toluene.
5. What is the thermal conductivity of tantalum?
6. Define the Stefan–Boltzmann law.

CHAPTER 4

Documents: Patents, Theses and Dissertations, Reports, etc.

Dewey 530.4, 530.8 Classes, etc.

IN THIS chapter the aim is to indicate what sort of information is available from documents, patents, theses and dissertations, reports and other published material excepting periodicals and books. The information in documents may never find its way into a periodical or a textbook and so special searches need to be made.

Patents

A patent is defined as "a monopoly or right granted according to U.S. patent law or British patent law or under similar statutes for the protection of inventions or discoveries". The patent specification sets out precisely the limits of the monopoly claimed.

Patents are granted to individuals or companies, who in return for the granting of the monopoly for a limited time, disclose their invention. Under British patent law it is "a grant by the Sovereign that gives the true and first inventor or certain persons claiming under him, the right to exclude for sixteen years, with certain rights of extension, others from the manufacture or use of the inventor's commercially vendible, original and useful article or method or process of manufacture or of control, improvement or modification thereof or of any such new and useful method or process of testing such manufacture, control or improvement that embraces any substance or material and any plant, machinery or apparatus, and that is sometimes subject in the public interest to compulsory licenses or to revocation".

The procedure of applying for a patent is simply outlined by A. C. Smith in a chapter in *Use of the Chemical Literature*, Butterworth, London, 1962. The title of a patent generally gives very little information about the nature of the patent and one must look at the body of the patent for detailed information, especially the section dealing with the scope of the monopoly claimed. British patents have the following general pattern:

General background.
Precise statement of invention.
Explanation of factors involved.
Specific working examples of the invention.
The formal claims—these are essentially legal.

When examining a patent application, the Patent Office examiner searches published literature for prior publications on the invention and the results of this search are available in Britain if one files a form, together with a fee, at the Patent Office, but in the case of German and United States patents a list of references, found in the search by the Patent Office, is given in the patent.

To keep abreast of current patent literature one must either subscribe to one of the commercial services available, or search personally the official journals and patents.

The Patent Offices of the major countries issue official journals, e.g. *Official Journal*, Patent Office, London; *Official Gazette*, U.S. Patent Office, Washington D.C. These list recent patent applications and complete specifications. By scanning them one can keep up to date with the patent literature.

To do retrospective searching for patents, it is necessary to consult the published keys issued by the various Patent Offices.

Theses, Dissertations

Bodies issuing lists of theses and dissertations are discussed by I. R. Stephens in an interesting article which is included in *Searching the Chemical Literature*, American Chemical Society, Washington D.C., 1961. The countries covered are Australia,

Austria, Canada, Denmark, France, Germany, Great Britain, India, Netherlands, Russia, South Africa, Sweden, Switzerland and the United States.

Aslib (3 Belgrave Square, London, S.W.1) publishes a yearly list—*Index to Theses Accepted for Higher Degrees in the Universities of Great Britain and Ireland*.

A survey of theses in British libraries was published in 1950, this being by P. D. Record, *A Survey of Thesis Literature in British Libraries*, Library Association, London, 1950. A guide to bibliographies of theses in the United States and Canada by T. R. Palfrey and H. E. Coleman is *Guide to Bibliographies of Theses, United States and Canada*, American Library Association, Chicago, 1940. *Dissertation Abstracts* lists theses originating from American universities and microfilms of the complete thesis are available from University Microfilms, Ann Arbor, Michigan, or their local agents. In the case of the U.K., the address is 44 Great Queen Street, London, W.C. 2. Many theses which are not microfilmed are included in the *Index of American Doctoral Dissertations*. Other published lists for the U.S.A. include *Masters Theses in Pure and Applied Sciences*, Thermophysical Properties Research Center, Purdue University, and M. L. Marckworth's *Dissertations in Physics*, Stanford University Press, Stanford, California, 1961, which contains an index to theses from 1861 to 1959; *Doctoral Dissertations Accepted by American Universities*, Wilson, New York, 1933 to date; *List of American Doctoral Dissertations Printed in 1912 to 1938*, U.S. Government Printing Office, Washington D.C., 1913–40. 26 volumes; *Doctorates Confirmed in the Sciences by American Universities*, National Council Report and Circular Series Nos. 12, 26, 42, 75, 80, 86, 91, 95, 101, 104, 145.

Several indexes to theses in other countries are published and these include *Deutsche Bibliographie—Jahresregister zur Bibliographie Deutschland–Oestereich–Schweiz*, Buchhändler-Vereinigung, Frankfurt-am-Main. A subject and author index which includes West German, Austrian and Swiss theses. *Jahresverzeichnis der Deutschen Hochschulschriften*, Deutsche Bücherei,

Leipzig, covers German theses. *Union List of Higher Degree Theses in Australian University Libraries*, University of Tasmania Library, covers theses held in particular Australian University libraries. *Catalogue des Thèses et Écrits Académiques*, Ministère de l'Éducation Nationale, Paris, covers French theses and they are also listed in *Bibliographie de la France*.

Reports

When work is being done at a laboratory it is usual for progress reports to be written at intervals and for a final report to be issued when the work is completed. This final report will be circulated to those within the laboratory having interest in the work, and, depending on the commercial or security aspects, to others not within the laboratory or company. The report may also be published, either in full or in an abbreviated form, in a learned journal, or it may be available for sale from various agencies. Of particular interest are the reports of the various centres working on atomic energy throughout the world. In most cases, if a report is freely available, details will be found in due course in *Nuclear Science Abstracts*. From this publication one can also trace the libraries which hold reports from the various agencies and also where to buy these reports. The United Kingdom Atomic Energy Authority's reports available for purchase are listed in the HMSO's *Daily List of Government Publications* and in a useful monthly list issued by the UKAEA, Harwell, Berkshire— *U.K.A.E.A. List of Publications Available to the General Public*.

Guides to UKAEA publications include the following:

ANTHONY, L. J., *Sources of Information in Atomic Energy*, HMSO, London, 1960.

SMITH, J. R. (ed.), *Guide to U.K.A.E.A. Documents*, HMSO, London, 2nd ed., 1960. This covers UKAEA information and its availability, UKAEA document series and referencing systems and UKAEA unclassified bibliographies.

Reports from other U.K. government laboratories which can

be purchased are listed in the *Daily List of Government Publications*, HMSO, London, whilst reports from U.S. governmental and civilian agency laboratories are found in *U.S. Government Research Reports*, U.S. Department of Commerce, Washington D.C.

Other Documents

This heading is used as a general term for annual reports of laboratories, company annual reports, trade directories, etc. The National Physical Laboratory, Teddington, Middlesex, and the National Institute for Research in Nuclear Science, Harwell, Berkshire, are amongst the many governmental and semi-governmental laboratories issuing annual reports from which information on the work done during the previous year can be obtained. These are listed by HMSO in the *Daily List of Government Publications* or in the monthly cumulation *Government Publications*, or in sectional lists of the publications of particular departments, e.g. *Sectional List No. 3—D.S.I.R., Sectional List No. 63—Atomic Energy*.

A brief guide to British governmental publications was issued in 1960 by the Stationery Office. An earlier useful publication is R. Staveley's *Government Information and the Research Worker*, Library Association, London, 1952. A comprehensive guide to U.S. governmental publications is L. F. Schneckebier's and R. B. Eastin's *Government Publications and their Use*, Brookings Institution, Washington D.C., 1961. The monthly list of U.S. governmental publications is the *United States Government Publications Monthly Catalog*. Another useful guide which lays emphasis on British, U.S. and international organizations is E. S. Brown's *Manual of Government Publications, United States and Foreign*, Appleton Century, Crofts, New York, 1950.

Non-governmental laboratories such as the Arthur D. Little Laboratories issue either annual reports or occasional brochures describing the type of work they are capable of doing and quoting examples.

Trade literature of individual manufacturers may detail new

equipment or processes of value to the physicist. Literature of this sort will probably be filed in the library in an alphabetical sequence under the name of the company, with possibly a subject index.

The various trade associations usually issue catalogues giving the names of member firms and their products. Of particular interest in this category is one issued by the Scientific Instrument Manufacturers Association, *British Instruments Directory and Buyers Guide*, United Science Press, London. This includes an alphabetical list of products and manufacturers, whilst a similar publication covering electronics is *Communications and Electronics Buyers Guide and Who's Who*, Heywood, London.

For documents relating to specialized aspects, see the appropriate chapter.

Questions

1. What sort of information can be found from patents?
2. What source could be consulted for information on a thesis originating from Sweden?
3. How can recent reports originating from the National Physical Laboratory be traced?
4. Give details of some of the most recent reports of the French Commissariat a l'Énergie Atomique (CEA).
5. What is the programme of work being undertaken at the Arthur D. Little Laboratories in either the U.K. or the U.S.A.?

CHAPTER 5

Periodicals: General

Dewey 530.5 Class, etc.

THERE are now well over 100 periodicals devoted exclusively to physics and at least an additional 500 which contain some material of interest to physicists. *Physics Abstracts* in fact searches over 700 periodicals for its items. The growth of scientific literature is well illustrated by the number of entries in each of the editions of the *World List of Scientific Periodicals*. 25,000 entries were in the 1925–27 edition, 36,000 entries in the 1934 edition, whilst the 1952 edition covering the period 1900–50 has over 50,000 entries.

It has become increasingly difficult to keep up with the output of scientific periodicals. It has been estimated that it would take ten years to read all the literature published in one year. More and more journals have overlapping fields of interest, and some seem to be prepared to publish any material regardless of its worth. Solutions to this problem have been put forward, such as publishing only abstracts of papers and requiring those interested to contact the author for the full version, but, so far, no practical solution to the problem is in operation.

The very great advantage of periodicals over books is their frequency of publication, which means that the information in them is much more up to date than that appearing in textbooks, for it usually takes less time to write and publish a periodical article than a book. Thus the periodical is the place where details are published of the most recent research work, new processes and the latest developments in highly specific fields. Information such as market prices, manufacturers' announcements, news items, etc., can also be obtained from periodicals, and this is rarely summarized in abstracting publications. Hence physicists, like other scientists, can never rely entirely on abstracting publications.

For background information on the early growth of periodical literature, reference should be made to D. A. Kronick's *A History of Scientific and Technical Periodicals: the Origins and Development of the Scientific and Technological Press, 1665–1790*, Scarecrow Press, New York, 1962, and to a chapter on the growth of scientific periodical literature in a book by J. L. Thornton and R. I. J. Tully—*Scientific Books, Libraries and Collectors*, Library Association, London, 2nd ed., 1962.

Lists of periodicals of value to the physicist include:

A List of British Scientific Publications Reporting Original Work or Critical Reviews, Royal Society, London, 1950.

BRAY, R. S., *List of Periodicals of Physics Interest*, U.S. Department of Commerce, Office of Technical Services, Washington D.C., 1950. This list also indicates whether a periodical is abstracted by *Physics Abstracts*, *Chemical Abstracts* or *Nuclear Science Abstracts*.

British Union Catalogue of Periodicals, Butterworth, London, 1955–58, is an alphabetical record of periodicals of the world held in British libraries. More than 140,000 titles are listed.

Directory of Canadian Scientific and Technical Periodicals, National Research Council, Ottawa, 1961.

Directory of Japanese Scientific Periodicals, National Diet Library, Tokyo, 1962. More than 2200 titles are listed with details of frequency of publication, publisher, year of first issue, etc.

Guide to Current British Periodicals, Library Association, London, 1962.

Guide to Latin American Scientific and Technical Periodicals: an Annotated List, Pan American Union, Washington D.C., 1962. The guide is arranged by subject, physics being represented by twelve current periodicals.

Letopis Periodicheskikh Izdanii SSSR, Vsesoyuznaya Khniznaya Palata, Moscow (yearly). This is divided into two series: (1) new periodicals and newspapers, those which have changed their titles and those which have ceased to appear, (2) scientific papers, studies, monographs and collections classified by main subject.

Nifor Guide to Indian Periodicals, 1955–1956, National Information Service, Poona, 1955.

SMITS, R., *Serial Publications of the Soviet Union, 1939–1957: a Bibliographical Check List*, Library of Congress, Washington D.C., 1958.

Sperling's Zeitschriften und Zeitungs Addressbuch, Börsenverein der Deutschen Buchhändler, Leipzig.

The World List of Scientific Periodicals Published in the Years 1900–1950, Butterworth, London, 3rd ed., 1952. This lists about 50,000 scientific periodicals and indicates where they are available for consultation.

Ulrich's Periodicals Directory, Bowker, New York, 10th ed., 1963. Gives a selection of over 20,000 periodicals covering particular topics. For physics about 100 periodicals are listed, showing publisher, frequency of publication, whether reviews, bibliographies or abstracts are carried, and whether it is abstracted or indexed in any other publication.

WEISPEC, J., *Polish Periodicals, 1953–1956: an Annotated Bibliography*, Catholic University of America, Washington D.C., 1957.

Willings Press Guide, Willing, London (annually). This covers U.K. publications and the principal British Commonwealth, Dominion, Colonial and foreign publications.

The Battelle Memorial Institute has just completed *A Guide to East European Scientific and Technical Literature*, National Science Foundation, Washington D.C., 1963, which includes general statements on the publishing of scientific and technical literature in Albania, Bulgaria, Czechoslovakia, Hungary, Poland, Rumania and Yugoslavia. Information is given on the announcement, availability, procurement, exchange and translation of the publications noted.

Two publications which are available from the National Science Foundation in Washington are valuable for giving guidance on

Soviet literature and these are *Providing U.S. Scientists with Soviet Scientific Information* and *Russian Scientific Journals Available in English*. The former gives a useful account of Soviet scientific publications and gives details of those which are translated into English, whilst the latter gives the information in pamphlet form on translated journals.

Catalogues of early scientific journals also exist such as:

BOLTON, H. C., *A Catalogue of Scientific and Technical Periodicals, 1665–1895, together with Chronological Tables and a Library Check List*, Smithsonian Institute, Washington D.C., 2nd ed., 1897. A subject index is included, which, of course, very much adds to the usefulness of the catalogue. Excluded are transactions of learned societies, unless forming part of a series which had begun as, or later became, an independent periodical.

SCUDDER, S. H., *Catalogue of Scientific Serials of all Countries including the Transactions of Learned Societies in the Natural, Physical and Mathematical Sciences, 1633–1876*, Harvard University Library, Cambridge, Massachusetts, 1879. Arrangement is by country and town of origin. Less complete but complementary to Bolton's work referred to previously.

The most convenient lists of current periodicals are those of the abstracting services, such as *Physics Abstracts*, which indicate their coverage. These lists are to be found with the indexes to the abstracts. They are usually arranged in alphabetical order of the abbreviated titles used in the abstracts and give the full title of the journal and the name and address of the publisher.

Many of the most authoritative periodicals are published by societies.

The American Institute of Physics and its member societies issue many periodicals including:

American Journal of Physics. Stresses the educational, historical and philosophic aspects of physics. Issued monthly.

Applied Optics. Original papers in applied optics and related fields. Issued monthly.

Applied Physics Letters. Similar in aim, size and format to *Physical Review Letters* for reports of important new findings in applied physics. Issued monthly.

Astronomical Journal. Publishes original work in astronomy. Issued ten times a year.

Bulletin of the American Physical Society. Abstracts of papers given at all Physical Society meetings, minutes, advance information of meetings. Issued seven times a year.

Journal of the Acoustical Society of America. Original papers on all aspects of acoustics. Issued monthly.

Journal of Applied Physics. Applications of physics in industry and other sciences. Issued monthly.

Journal of Chemical Physics. Bridges the gap between physics and chemistry. Issued semi-monthly.

Journal of Mathematical Physics. Advances in mathematical techniques applicable to various branches of modern physics. Issued monthly.

Journal of the Optical Society of America. Original papers on all aspects of optics. Issued monthly.

Physics of Fluids. Original research in structure, dynamics and general physics of gases, liquids and plasmas. Issued monthly.

Physical Review. Reports original research in experimental and theoretical physics. Issued semi-monthly.

Physical Review Letters. Timely short reports of important new findings. By keeping the reports short this allows speedy publication. A further account usually appears later in *Physical Review.* Issued semi-monthly.

Physics Today. Developments in physics for physicists and others. Issued monthly.

Reviews of Modern Physics. Discussions of current problems of physics. Issued quarterly.

Review of Scientific Instruments. Covers scientific instruments, apparatus and techniques. Issued monthly.

Sound, its Uses and Control. Practical information on measurement and control of noise, shock, vibration and sound. Issued bi-monthly.

The American Institute of Physics is also responsible for the translation of eight Soviet physics journals. These are:

Soviet Astronomy (Astronomicheskii Zhurnal). Astrophysics and radioastronomy, stellar activity and instrumentation.

Soviet Physics—Acoustics (Akusticheskii Zhurnal). Physical acoustics but also includes electro-, bio- and psycho-acoustics. Mathematical and experimental work, chiefly pure research.

Soviet Physics—Crystallography (Kristallografiya). Theory and experiment on crystal structure, lattice theory, diffraction studies.

Soviet Physics—Doklady (The Physics Section of Doklady Akademii Nauk SSSR). Proceedings of the Academy of Sciences which reports recent research in physics.

Soviet Physics—J.E.T.P. (Zhurnal Eksperimentals noi i Teoreticheskoi Fiziki). This is the leading Soviet physics periodical and is similar to *The Physical Review* both in range of topics and quality.

Soviet Physics—Solid State (Fizika Tverdogo Tela). Experiments and theoretical investigations on dielectrics and semi-conductors and the applied physics associated with these problems.

Soviet Physics—Technical Physics (Zhurnal Tekhnicheskoi Fiziki). Plasma physics and magnetohydrodynamics, aerodynamics, ion and electron optics, mathematical physics, the physics of accelerators and molecular physics.

Soviet Physics—Uspekhi (Uspekhi Fizicheskikh Nauk). Comparable to *Reviews of Modern Physics*. Reviews recent developments, reports on scientific meetings and does book reviews.

The Optical Society of America is responsible for publishing a translation of *Optics and Spectroscopy (Optikai Spektroskopiya)*.

In Great Britain the periodicals issued by societies and institutions include:

British Journal of Applied Physics. Contains original work, reviews, technical news, review lectures, book reviews and correspondence on all branches of applied physics and especially on applications in industry. Issued monthly by the Institute of Physics and the Physical Society.

Discussions of the Faraday Society. Original work in physical chemistry, chemical physics, biophysics and colloidal chemistry. Issued half-yearly.

Journal of Scientific Instruments. Original work, reviews, historical notes, technical news, book reviews and correspondence on the principles, construction and use of scientific instruments. Issued monthly by the Institute of Physics and the Physical Society.

Philosophical Transactions of the Royal Society, *Series A.* Original work in physical and mathematical sciences. Issued irregularly.

Proceedings of the Physical Society. Contains original work, abstracts and book reviews. Section A is mainly microscopic physics and Section B mainly macroscopic. Issued monthly.

Proceedings of the Royal Institution. Original work, review lectures and reviews in general science and art. Issued three times a year.

Proceedings of the Royal Society. Series A. Original work, review lectures and reports of discussion meetings in mathematical and physical sciences. Issued irregularly.

Proceedings of the Royal Society of Edinburgh. Section A. Mathematical and Physical Sciences. Original work in mathematical and physical sciences. Issued irregularly.

Transactions of the Faraday Society. Original work and book reviews in physical chemistry, ᵇemical physics, biophysics and colloidal chemistry. Issued monthly.

Transactions of the Royal Society of Edinburgh. Original work in general science. Issued irregularly.

Other periodicals of physics interest issued throughout the world include:

Académie des Sciences, Paris, Comptes Rendus. Covers general scientific topics. Issued weekly.

Acta Physica Austrica. General physical topics. Semi-annual.

Acta Physica Polonica. All aspects of physics. Issued monthly.

Advances in Physics. General coverage of advances in particular physics fields. Issued quarterly as a supplement to the *Philosophical Magazine.*

Anales de la Real Sociedad Espanola de Fisica y Quinica, Seria A. Fisica, Madrid. General physics topics. Irregular.

Analis de Academia Brasileira de Ciencias. Original research in physical, natural and exact sciences. Issued quarterly.

Annalen der Physik. Covers all aspects of physics. Issued eight times a year.

Annales de Physique, Paris. Covers all aspects of physics. Issued bi-monthly.

Arkiv für Fysik, Stockholm. Broad field of physics and especially atomic physics. Issued irregularly.

Australian Journal of Physics. Covers all aspects of physics. Issued quarterly.

Cahiers de Physique. Broad field of physics, especially atomic physics. Monthly.

Canadian Journal of Physics. Original research. Issued monthly.

Contemporary Physics. A journal of interpretation and review. Issued bi-monthly.

Czechoslovak Journal of Physics. General physics topics. Issued monthly.

Fortschritte der Physik. Broad field of physics, especially atomic physics. Issued monthly.

Geophysical Journal. All aspects of geophysics. Issued bi-monthly.

Helvetica Physica Acta. All aspects of physics. Issued six to eight times per year.

Indian Journal of Pure and Applied Physics. All aspects of physics. Issued monthly.

Indian Journal of Theoretical Physics. All aspects of physics. Issued quarterly.

International Science and Technology. Covers all scientific fields. The articles are written so that one scientist can introduce his field to other scientists. Issued monthly.

Journal de Chimie Physique. Covers all aspects of physics including atomic and biophysics. Issued monthly.

Journal de Physique et Radium. All aspects of physics, especially atomic physics. Issued monthly.

Journal of Atmospheric and Terrestrial Physics. General coverage of the field. Issued monthly.

Journal of the Franklin Institute. All aspects of physics. Issued monthly.

Journal of the Mechanics and Physics of Solids. All aspects are covered. Issued quarterly.

Journal of the Physical Society of Japan. Original papers in physics and related fields. Issued monthly.

Nature. Original work, professional news and book reviews in general science. Issued weekly.

Naturwissenschaften. Covers general scientific topics. Issued twice monthly.

New Scientist. General scientific topics for the layman and the scientist. Issued weekly.

Pakistan Journal of Science. Covers the broad field of science. Issued bi-monthly.

Philosophical Magazine. Theoretical, experimental and applied physics. Issued monthly.

Physica. Covers original research in all fields of physics. Issued monthly.

Physica Status Solidi. Original research papers in the field of solid state physics. Issued monthly.

Physikalische Blätter. General coverage of all physical topics. Issued monthly.

Physics Express. This is a comprehensive digest of current Russian literature dealing with physics topics and is issued by International Physical Index, 1909 Park Avenue, New York. Each issue contains a sampling of articles selected from the totality of the latest published Soviet journals. Many of the articles are completely translated, some are excerpted or summarized and others abstracted, depending upon relative importance, time liness and content. The major divisions include: atomic structure and spectra; cosmic rays, elementary particles and accelerators; cryogenics and superconductivity; dielectrics and ferroelectricity; magnetism, ferri- and ferromagnetic phenomena; mathematical methods; molecular structure and spectra, nuclear structure; physics of space and geophysics; scattering; solid state and semiconductors; atomic energy; beams and plasmas.

Proceedings of the National Academy of Sciences of the U.S.A. Publishes promptly brief first announcements of the results of original research by members of the Academy. Issued monthly.

Proceedings of the National Institute of Sciences of India: Pt. A. Covers physical sciences generally. Issued bi-monthly.

Revista de Fisica. Descriptive and survey articles, translations, class and course notes. Issued quarterly.

Revista Mexicana de Fisica. Original research in the broad field of physics and especially atomic physics. Issued quarterly.

Revue de Physique (Roumaine). Covers general physics. About six issues per year.

Science Record. Published in English and aims at keeping the scientific world informed of the development of scientific research in China. Articles are written by Chinese scientists.

Scientia Sinica. Published in English and is devoted to publishing articles selected from various Chinese periodicals.

Scientific American. General science for the scientist. Issued monthly.

South African Journal of Science. Covers the broad field of science. Issued monthly.

Zeitschrift für Angewandte Mathematik und Physik, Basle. Covers all aspects of applied mathematics and physics. Issued bi-monthly.

Zeitschrift für Angewandte Physik, Berlin. Original papers in all fields of applied physics with some comprehensive papers covering several fields. Issued monthly.

Zeitschrift für Physik, Berlin. General physics topics. Issued monthly.

Most of the leading newspapers carry a scientific correspondent on its staff who reports latest developments in many fields and these can be very valuable sources of information. Particularly valuable in this context is the *Financial Times*.

Translations of Periodicals

It is becoming increasingly important, if one is intent on keeping up to date, to keep informed of the work being done in foreign countries. Consequently, many translations are published.

Lists of their translations are published by commercial organizations, such as Associated Technical Services, New Jersey; Consultants Bureau, New York, whose *Soviet Science in Translation* gives advance listing of all articles in 83 English versions of Russian journals; and Pergamon Press, London, etc.

In the United States, the Office of Technical Services, U.S. Department of Commerce, Washington D.C., collects translations from Federal agencies and foreign sources, and the Special Libraries Translation Center, John Crerar Library, Chicago, collects translations from non-governmental sources.

American guides to translations include:

Bibliography of Translations from Russian Scientific and Technical Literature, Library of Congress, Washington D.C., 1953–56.
Translations Monthly, John Crerar Library, Chicago, 1955–58.
Technical Translations, U.S. Government Printing Office, Washington D.C., 1959–.

In Britain, the National Lending Library (NLL), Boston Spa, Yorkshire, is responsible for providing literature in all fields of science and technology and for translation into English from difficult languages. Its collection of Russian translations includes over 20,000 articles and 500 books. The NLL publications include: *N.L.L. Translations Bulletin, List of Books Received from the U.S.S.R. and Translated Books* and *Titles of Periodicals from the U.S.S.R. and Cover to Cover Translations*.

Aslib, Belgrave Square, London, S.W. 1, maintains the *Com-*

monwealth Index to Unpublished Translations which covers translations made in all countries in the Commonwealth. Other bodies which may undertake translations in their own field include the United Kingdom Atomic Energy Authority, the Royal Aircraft Establishment, the Building Research Station, the Chemical Society, the British Transport Commission, the Rubber and Plastics Research Association, the Production Engineering Research Association, the British Iron and Steel Translation Service, the Physical Society and the Institute of Physics.

European sources of translations include the Auswertungsstelle für Russische Literatur, Technical University, Hanover. A complete file of translations made in Eastern Germany is kept by the Institut für Dokumentation, Academy of Sciences, Berlin. The European Productivity Agency has established a European Translations Centre at Delft to publicize existing translations into Western languages of Russian and Eastern European scientific literature. It issues *Russian Technical Literature* which lists translations available, in progress or planned. *Transatom Bulletin* lists translations covering nuclear literature, in particular from the less familiar languages such as Japanese. It is issued by Transatom, a co-operative effort in Brussels of the European Atomic Energy Community, U.S. Atomic Energy Commission and the United Kingdom Atomic Energy Authority. UNESCO's *Index Translationum* is a general international index of translations. This is issued quarterly; it is arranged by country, and subdivided by subject. A good survey of organizations for translating and lending translations is *Scientific and Technical Translating*, UNESCO, Paris, 1958.

For periodicals concerned with specialized topics, see the appropriate chapter.

Questions

1. Why are periodicals so important?
2. Compare *Physical Review* and *Soviet Physics—J.E.T.P.* (*Zhurnal Eksperimentals noi i Teoreticheskoi Fiziki*).
3. Has a translation been published of T. I. Filippova and N. V. Filippov, "Measurement of the electron temperature of the plasma in a strong shock wave," *Yadernyi Sintez*, **1**, 195 (1961)?
4. What periodicals published in Japan are of interest to the physicist?
5. How many of the periodicals found in answer to question 4 are abstracted in *Physics Abstracts*?

CHAPTER 6

Abstracts: General

Dewey 530.5 Class, etc.

AN ABSTRACT has been defined as "a summary of a publication or article accompanied by an adequate bibliographical description to enable the publication or article to be traced". Generally, an abstract may be indicative or informative. An indicative abstract is usually short, picking out some highlights from the article so that the reader can decide whether or not to read the original. An informative abstract presents data given in the original article and summarizes the methods and arguments used.

When searching for information in abstract publications one should start from the present date and work backwards towards the earliest publications. Before starting to search, the subject index should be scanned and all the headings which may be used noted, so that no information is overlooked.

At the UNESCO Conference on Scientific Abstracting in Paris in 1949, a proposal for the formation of a single physics abstracting journal was referred to a Committee of the International Union of Physics. This Committee concluded that, though this proposal was an attractive one, it was necessary for abstracting journals to be available in three or four different languages. The indexes must be in a language familiar to the searcher, since the value of the entry might depend on a single word. An interesting report of the co-operation existing between some of the abstracting organizations noted that there had been no move by the abstracting organizations themselves to share the actual work of abstracting. It pointed out that an abstracting organization must work systematically as a fairly closely knit team and that, when the difference in the basis and standards of selection, appraisal, classification, types of abstract and indexing between each organization is considered, it is not surprising that each organization preferred to abstract independently.

Only about half the papers given at conferences are published in full and are available as abstracts. The policy of *Science Abstracts* is not to publish abstracts of papers presented orally at meetings, unless précis of substantial length are published in the normal way. The Institute of Physics publishes summarized proceedings of conferences organized by groups of the Institute in its periodicals, but the names of the authors of the papers are not indexed. Some conference proceedings are published in full as supplements to periodicals.

It has been suggested that the quality of abstracts may be partly indicated by the name of the abstractor. Hence some abstracts, such as those in *Applied Mechanics Reviews*, are signed with the full name and country of the abstractor, while others, as in *Chemical Abstracts*, are signed with the full name or initials of the abstractor. *Nuclear Science Abstracts* and *Physics Abstracts* carry some signed abstracts with a large number of author's abstracts. Staff writers compile the abstracts in the case of *Engineering Index*, and these are uniform but descriptive rather than analytical. The author's or contractor's abstracts in U.S. Government Research Reports are uniform and adequate.

Nuclear Science Abstracts, *Physics Abstracts* and *Applied Mechanics Reviews* include a subject index in every issue. *Engineering Index* is divided alphabetically into subject fields and has recently become available monthly.

Often more than one abstract periodical must be searched for completeness. On one occasion when the work of P. C. Thonemann was being studied, a list of eight periodical articles, six letters to the editor and one research report was compiled and a search was made for abstracts of them. Six items were abstracted in *Nuclear Science Abstracts*, 1948–15 December 1956, five in *Physics Abstracts*, 1944–November 1956 and four only in *Chemical Abstracts*, 1943–November 1956.

A survey in 1950 showed that the abstracting services used by most physicists were:

Physics Abstracts (*Science Abstracts A*)	Used by 93% of 1477 physicists interviewed.
Chemical Abstracts	Used by 40% of 1477 physicists interviewed.
Nuclear Science Abstracts	Used by 28% of 1477 physicists interviewed.
Electrical Engineering Abstracts (*Science Abstracts B*)	Used by 18% of 1477 physicists interviewed.
Mathematical Reviews	Used by 10% of 1477 physicists interviewed.
Applied Mechanics Reviews	Used by 7% of 1477 physicists interviewed.
Engineering Index	Used by 7% of 1477 physicists interviewed.

In order of priority, these physicists looked for wide coverage, prompt publication, technically qualified abstractors, extensive indexes and maintenance of the cost below $5 per year.

Guides to general abstract journals and bibliographies of value to the physicist include:

A Guide to the World's Abstracting and Indexing Services in Science and Technology, National Federation of Science Abstracting and Indexing Services, Washington D.C., 1963.

A World Bibliography of Bibliographies, Societas Bibliographica, Geneva, 3rd ed., 1955–56, compiled by Theodore Besterman, indexes over 84,000 bibliographies by subject.

Bibliographic Index, issued quarterly by H. W. Wilson. It lists separately published bibliographies as well as contributions to books, periodicals and serials.

Bibliographical Services throughout the World 1950–1959, UNESCO, Paris, 1961, compiled by R. L. Collison.

DARROW, K., *Classified List of Published Bibliographies in Physics, 1910–1922*, National Research Council, Washington D.C., 1924.

DOROSH, J. T., *Guide to Soviet Bibliographies*, Library of Congress, Washington D.C., 1950.

Index Bibliographicus, 4th ed., which includes 1855 titles from 50 countries and an alphabetical subject index.

List of Current Abstracting and Indexing Services, FID, The Hague, 1949.

TANGHE, R., *Canadian Bibliographies*, University of Toronto Press, Toronto, 1960.

Technical Information in the U.S.S.R., Massachusetts Institute of Technology, Cambridge, Massachusetts, 1961. (A translation of the 1960 Russian edition by A. S. Melik.)

Specialized or regional abstracting or indexing services of value to the physicist include:

Applied Mechanics Reviews. Covers such fields as heat, fluid and solid mechanics, gas dynamics, etc., and includes books, congresses, journals, patents, published reports, published translations, dissertations, and is very selective on unpublished material. Abstracts are informative, including critical reviews and written by subject specialists. *Applied Mechanics Reviews* has been published monthly since 1948 by the American Society of Mechanical Engineers.

Applied Science and Technology Index. Covers all fields with entries under subject headings. No abstracts are given. It is issued monthly with quarterly and yearly cumulations. The average time between publication of an article and inclusion in the index is about three months. Publication under this title commenced in 1958, previously it had been part of the *Industrial Arts Index* which has been published by H. W. Wilson, New York, since 1913.

Australia Science Index, CSIRO, Melbourne. Covers all aspects of pure and applied science. Twelve parts are published each year containing in total about 3600 references. A yearly index is issued.

Bibliografia Brasileira de Matematica e Fisica. Published by the Instituto de Bibliografia e Documentacao, Av. General Justo

171, Rio de Janeiro. Indexes forty-two mathematics and physics periodicals, twelve of which are Brazilian, and includes books and theses. Issued irregularly.

Bibliographie Scientifique Française. Published by Gauthier-Villars, Paris. Covers the period from 1902.

Bibliography of Scientific Publications of South and South-East Asia. This is published monthly and the entries are classified. Originally it was issued in 1949 by the UNESCO Science Co-operation Office in New Delhi as a half-yearly publication. From 1955, Insdoc have been associated with the work and in 1956 and 1957 it was issued quarterly.

Bilten Dokumentacija Struene Literature. Published monthly in six parts by the Yugoslovenski Centur za Technicka i Naucau Dokumentaciju, Belgrade. Each part contains short abstracts in Serbo-Croatian arranged by UDC. Approximately 40,000 abstracts are published each year.

Boletin del Centro de Documentation Cientifica y Technica. Published monthly in five parts by the Centro de Documentacion Cientifica y Technica de Mexico, Enrico Martinez 24, Mexico. Section 1 deals with mathematics, astronomy and astrophysics, physics, geology, geophysics and geodetics and lists all substantive articles included in journals recently received by the Centre. In 1959, the Centre received 2895 periodicals regularly and its collection of Latin American periodicals is the most comprehensive single collection in existence. The title of the article is given in the original language with a Spanish translation if needed and articles published in Latin American countries are indexed with an asterisk to draw immediate attention to them. Abstracts of articles published in Latin American periodicals are prepared by subject specialists and published in English and Spanish. A detailed subject index was to be added in 1961.

British Abstracts. Ceased publication in 1953. The periodical coverage of the chemical literature, with some abstracts of interest to the physicist, was fairly comprehensive for papers quoting original research and for review articles. Subject experts wrote the abstracts which were indicative and without critical review. Patents and government publications were covered. Section A1—general, physical and organic chemistry, carried about 2435 abstracts in 1948. Section B1—chemical engineering and industrial inorganic engineering, carried about 5751 abstracts in 1948. Section C—analysis and apparatus, carried about 2513 abstracts in 1948. About 1600 periodicals were covered. Indexes for author, subject and formulae were issued annually and every five years. Publication was monthly by the Bureau of Abstracts, 9–10 Savile Row, London, W.1. The earlier titles of this publication were 1924–25, *Abstracts of Chemical Papers. A: Pure Chemistry;* 1926–27, *British Chemical Abstracts;* 1938–44, *British Chemical and Physiological Abstracts.*

British Technology Index. The main subject fields covered are engineering, chemical technology, mining, metallurgy, metal manufacturers, wood manufacturers, textiles, clothing, paper-making, packaging, works management, industrial economics of particular industries, industrial health and safety, technical education. Entries are under subject headings, no abstracts being given. Only British periodicals are covered. Commenced publication in January 1962 and is published monthly by the Library Association, London.

Buletin de Documentare Technica. Published monthly by the Institute of Bibliography, Documentation and Technical Publishers, Street Gabriel, Peri No. 3, Bucharest, Rumania. Articles covered are those of interest and importance to Rumanian industry. Abstracts are partly informative and semi-informative, without critical review, and are written by subject experts. No indexes are published.

Bulletin Signalétique. Formerly known as the *Bulletin Analytique.* Published by the Centre de Documentation du Centre National de la Recherche Scientifique, Paris. First published in 1940, and since the beginning of 1961 it has been published in 22 monthly parts. Abstracts, which are usually of the indicative type, are grouped under broad subject headings. Each paper is noted in its original language with a translation of the title into French. An author index is contained in each issue, and an annual subject index is issued. In 1958 the editors estimated that the average time between the arrival of a periodical in a language other than French and it being noted in the abstracts was about six months. It contains about 270,000 abstracts per year and, in 1961, 9420 periodicals were received, a figure which the CNRS expects to increase to between 15,000 and 20,000 in 1963.

Centro di Documentazione Scientifico–Tecnica Consiglio Nazionale della Ricerche, Rome. Between 1949 and December 1955, it published a series of abstracts covering scientific and technical literature, both Italian and other, from journals received by the Centro di Documentazione. It was issued in fourteen sections, section 8 being the one dealing with physics.

Chemical Abstracts. Whereas *Physics Abstracts* generally limits its material to articles on physics research, *Chemical Abstracts* includes everything published that is chemical. Here chemical means "relating to the composition or structure of matter or changes therein" and thus includes much of physics. It is generally accepted as the most comprehensive journal of its kind, and covers books, congresses, periodicals, patents, published reports, published translations, dissertations and a certain amount of unpublished material. It is issued semi-monthly with abstracts being grouped under broad subject headings. An author index is included in each issue and starting in 1963 each one now contains a keyword subject index. An annual subject and author index is issued. Published by the American Chemical Society, it was first issued in 1907. To overcome delays, the American Chemical Society now also publishes *Chemical Titles* which is issued semi-monthly. The use of computers has made it possible to provide this "current awareness" service in the form of an index. The particular form in this concordance is the "keyword in context" index.

Current Contents. Published since January 1959 by the Institute of Scientific Information, Philadelphia, it is published in two

C

editions, one covering space and physical sciences. It lists titles in 500 journals, consisting of photographic reproductions of tables of contents.

Documentare Technica. Issued monthly since 1949 by the Institutal de Documentare Technica, Bucharest. Includes sections on technical and scientific works of general interest, energetics, electrical engineering, electronics, etc. Abstracts are grouped under broad subject headings, the titles of articles being given in the original language. Contents lists are published in Russian, English, French and German. About 2000 abstracts are listed annually.

Documentation Bulletin of the National Research Centre of Egypt, Cairo. Published monthly in two parts. Part I lists titles of papers received in the library, titles published in European languages and translated into English. Part II lists contents of periodicals published in Egypt and other Middle East countries. The titles are translated into English. A short index is published. The number of titles indexed in 1956 was 54,478.

Documentation Technique. Published monthly under the direction of the Centre de Documentation, 3 Rue de Messins, Paris. Comprehensive coverage of technical articles on all aspects of electricity. Conferences, standards and translations are included.

Electrical Engineering Abstracts (Science Abstracts B). Published monthly by the Institution of Electrical Engineers, Savoy Place, London, W.C. 1. Comprehensive coverage of publicly available literature contributing in some way to the advancement of electrical engineering practice. Abstracts are informative without critical review and are generally written by subject experts, the authors' abstracts being generally used for articles in the best journals. Abstracts are grouped under broad subject headings and each issue contains an author index. Annual indexes are issued.

Electronic Engineering Master Index. This service provides a bibliographical listing of research in electronics, optics, physics and allied fields. The listing is done under some 400 subject headings. Published by Electronics Research Publishing Company, 480 Canal Street, New York 31, N.Y. First edition covered the period 1925–45.

Engineering Index. Covers books, congresses, periodicals, published reports and some translations. It does not cover patents, dissertations or unpublished reports. Abstracts are semi-informative and without critical review and are written by its staff engineers. Indexing is by subject alphabetically with prolific cross-references. There is an author index. The 1950 edition had 1383 pages, the 1955 edition 1251 pages and the 1960 edition 1732 pages. Over 1400 periodicals are covered. Published by Engineering Index, 29 West 39th Street, New York 18, N.Y., it was published yearly, but there was also a subscription service available whereby abstract cards were issued very frequently. Towards the end of 1962, a monthly issue became available, thereby increasing its value.

English Abstracts of Selected Articles from Soviet Block and Mainland China Technical Journals. Issued by the Office of Technical Services, Washington D.C., it is a selective service covering about 200 Russian, Chinese and other less familiar technical journals.

Fuel Abstracts and Current Titles. Formerly *Fuel Abstracts.* Published monthly by the Institute of Fuel, Devonshire Street, London. It is a summary of world literature on all technical and scientific aspects of fuel and power, abstracts being grouped under broad subject headings. Each issue contains an author and subject index.

Hungarian Technical Abstracts. Published quarterly since 1948 in English, Russian and German by the Hungarian Central Technical Library and Centre for Documentation, Muzeum U 17, Budapest. It contains abstracts, which cover physics, electronics and mechanics, of the most outstanding articles featured in the publications of the Hungarian Academy of Sciences, of Hungarian Universities and in Hungarian technical and related periodicals, as well as abstracts of the publications of scientific institutes and of dissertations submitted to the

Hungarian Academy of Sciences by candidates for an academic degree. An author index is issued.

IBZ Internationale Bibliographie der Zeitschriftenliteratur. The first part indexes 3600 German language periodicals and 45 newspapers, the second part covers 3200 non-German titles, whilst the third part indexes German and non-German book reviews. Published semi-annually by Dietrich, Osnabruck.

Insdoc List of Current Scientific Literature. Issued twice a month by Insdoc (Indian National Scientific Documentation Centre), National Physical Laboratory, New Delhi, India. This is a classified list of titles of papers in various Indian and foreign periodicals. Its criteria is (a) comprehensiveness for the immediate needs of the country as far as facilities permitted and (b) a speedier supply of information than that available from existing abstracting or indexing periodicals. Titles in European languages are translated into English. No annual index is published.

Instruments Abstracts. Compiled by the British Scientific Instrument Research Association, Chislehurst, England and published monthly by Taylor & Francis, London. The subjects covered are astronomy, atomics and nucleonics, biology and medicine, chemistry, control, data handling, electricity and electronics, fluid mechanics, geophysics and meteorology, heat, light, magnetism and electromagnetism, materials and design, mechanics, navigation, surveying, telecommunications. Annual author and subject indexes are issued.

International Catalogue of Scientific Literature, 1901–1914. Published for the Royal Society of London by the International Council, London. Harrison, London, 1902–21. Fourteen annual issues, each in seventeen volumes. Each volume was devoted to a science. Volume A: Mathematics. Volume B: Mechanics. Volume C: Physics, etc. Items are classified according to the listed schedules and with author and subject approaches to book and periodical literature. Publication was suspended after completion of the volume for 1914. Covering the period 1800–1900 is the *Royal Society of London Catalogue of Scientific Papers, 1800–1900.*

Japan Science Review. Issued by the Scientific Information Division, Ministry of Education, Tokyo, in co-operation with other agencies. It is published in four series as a half-yearly publication. The series most likely to interest the physicist would probably be engineering sciences. These review and note the results of researches reported in Japanese periodicals. The average delay in 1957 for the circulation of information was about a year.

Letopis Zhurnalnykh Statei. (Chronicle of periodical articles.) A weekly Russian listing, under thirty-one main headings, of journal articles. About 2500 references are carried each week and about two-thirds of these concern science and technology.

Mathematical Reviews. Published monthly, except August, by the American Mathematical Society with editorial offices at Brown University, Providence 12, Rhode Island. The coverage is comprehensive for research papers in mathematics and mathematical physics. Abstracts are informative without critical reviews and as far as possible written by subject specialists. Annual author and subject indexes are produced.

Monthly Index of Russian Accessions. Published by the Library of Congress, Washington, it records all Russian language publications acquired by the Library and a group of about sixty cooperating libraries. The three main parts give (1) a list of monographs with an English translation of the title, (2) a list of contents of Russian periodicals with an English translation, (3) a subject index to the monographs and periodical articles.

Nuclear Engineering Abstracts. Published by Silver End Documentary Publications, London. The journal is addressed to the engineer engaged in nuclear work, and to the scientist employing nuclear instruments and machines. Abstracts are indicative and the coverage includes newspapers. Subject and author indexes are included in each issue.

Nuclear Science Abstracts. Covers books, congresses, periodicals, patents, published and some unpublished reports, published and some unpublished translations and some dissertations. Non U.S. atomic energy reports are included. It was originally published in 1946 under the title *Guide to Published Research on Atomic Energy*, this title being altered to *Abstracts of Declassified Documents* in 1947 and to *Nuclear Science Abstracts* in 1949. As its second title shows the original intention was to abstract U.S. reports and similar documents which had been released from security restriction as a result of the 1946 U.S. Atomic Energy Act. It is published fortnightly under the control of the U.S. Atomic Energy Commission, and each issue has a subject, author, corporate author and report serial number index. Cumulations are issued quarterly, half-yearly, yearly and quinquennially. A.E.C. research and development reports falling outside the scope of *Nuclear Science Abstracts* are abstracted in *Research and Development Abstracts of the U.S.A.E.C.*

Physics Abstracts. (*Science Abstracts A.*) Published in London with the support of the American Physical Society and the Institute of Physics and the Physical Society, *Physics Abstracts* covers books, congresses, periodicals and published reports; it does not cover patents, dissertations or much unpublished material. A broad subject index is included with each issue, although it cannot be adequate for all needs. The increase in the number of scientific publications concerning physics is reflected in the growth of *Physics Abstracts*. The 1950 issue had 9198 abstracts, 1955 had 10,160 and the 1960 issue had 21,407 abstracts. It has been estimated that it takes about three months for *Physics Abstracts* to abstract papers in English published in British periodicals, almost four months for papers in English from American periodicals, whilst it takes about six months to notice papers published in a European continental language. A survey showed that of 1477 physicists interviewed, 96% of the respondents rated it either good or medium on the quality of its abstracts and that the most frequent adverse comment was regarding the quality of the index. It is interesting to note that the American Institute for Physics is doing a survey of the *Physics Abstracts*. The points they are considering are (1) which journals are abstracted, (2) to what extent are books and conference proceedings covered, (3) how many months are needed to complete abstracting of classes of journals posing particular problems and (4) how abbreviations for journals in footnote references and library lists differ.

Physics Express. Published monthly since June 1958 by International Physics Index, New York. Russian language periodicals are searched and lengthy abstracts given in English. Mathematics, diagrams and photographs are reproduced. An author index is published.

Physikalische Berichte. Formerly *Die Fortschritte der Physik.* Issued monthly by Deutsche Physikalische Gesellschaften E.V., Braunschweig. Covers all phases of physics, mathematics, instrumentation, astrophysics, geophysics selectively. Abstracts which are indicative and informative are grouped under broad subject headings, the title of the article abstracted being in the original language. Monthly and annual author indexes are issued.

Polish Technical Abstracts. Published by Centralny Instytut Dokumentacji Naukowo-Techniczenj, Warsaw, Aleja Niepodleglosci 188. Published quarterly, first issued in 1951. Contains abstracts in English and Polish under broad subject headings such as mechanics, electrotechnics, power, which are taken from technical periodicals, books and pamphlets appearing in Poland.

Reportorium Commentationum a Societatibus Litteraris Editorum. J. D. Reuss, Gottingen, Apud Henricum Dietrich, 1801–21. Sixteen volumes. Volume 4 Physica (1805) covers society publications by classified subject index with an author index.

Revue Mensuelle des Sommaires des Principaux Périodiques Scientifiques et Techniques. The Centre de Documentation du Centre National de la Recherche Scientifique, Paris, issues this to inform interested parties of the content of the main periodicals and it reproduces tables of contents. Published on 35 mm microfilm as a monthly review. It covers 300 periodicals in such fields as physics, chemistry and biology.

Royal Society of London. Catalogue of Scientific Papers, 1800–

1900. Royal Society, London, 1867–1925. Nineteen volumes. Covers more than 1500 periodicals and transactions of learned societies and institutions. A subject index which covered physics, mechanics and mathematics was issued 1908–14. Covering the period 1901–14 is *International Catalogue of Scientific Literature*.

Russian Abstracts. This publication, which is now defunct, contained selected abstracts from some of the leading Russian technical periodicals and books. Periodicals for which "cover-to-cover" translations are published were not included. Publication started in 1958 and sixty-six issues in all were published, the last being issued in June 1961; each item listed carries an informative abstract. The body responsible for this publication was the Ministry of Aviation, Technical Information and Library Services.

Referativnyĭ Zhurnal. U.S.S.R. Institute for Scientific and Technical Information Abstract Journal which includes sections on: physics—monthly edition, each edition usually carrying about 2500 abstracts, annotations and bibliographical descriptions; geophysics—monthly edition, each edition usually carrying about 850 abstracts, annotations and bibliographical descriptions; electrotechnics—fortnightly edition, each edition usually carrying about 2100 to 2300 abstracts, annotations and bibliographical descriptions. *Mekhanika*. Published monthly and contains abstracts of articles from world scientific and technical literature. A translated version containing only Soviet literature is published by the Ministry of Aviation, Technical Information and Library Services TIL (c), First Avenue House, London, W.C. 1. The item numbers quoted are those that appear in the *Referativnyĭ Zhurnal*. The U.S. Office of Technical Services also issues a translated version of certain sections including physics.

It takes from four to six months from the moment the source arrives at the Institute to the time it is reflected in the Abstract Journal series. The following is a description of the abstracts in this series: "the abstract invariably gives the object, principal theoretical perquisites, method, results of the work, numerical data of scientific and technological interest, as well as the author's view on the possibilities for applying the results of the work to science and technology." Books, periodicals, dissertations and patents are included in the abstracts. The volume of information published in *Referativnyĭ Zhurnal* has increased to forty-two times its size since 1953. In 1960, almost 700,000 abstracts were noted.

Science Abstracts of China. Academia Sinica, Institute of Scientific Information, 117 Chao Yang Men Street, Peking, 1958 to date. Published in five separate sections: mathematical and physical sciences; chemistry and chemical technology; biological sciences; earth sciences; technical sciences. Each section is issued bi-monthly in English. Abstracts are grouped under subject headings.

Soviet Technology Digest. Express information on recent technological developments in the Soviet Union and Eastern Europe, consisting of illustrated abstracts from many journals and contents lists of books and periodicals. Published by Pergamon Press, London.

U.S. Government Research Reports. Formerly called *Bibliography of Scientific and Industrial Developments.* This publication comes from the U.S. Department of Commerce, Business and Defence Services Administration, Office of Technical Services, and has been published since 1933. It is issued twice a month to announce new reports of research and development released by the Army, Navy and Air Force, Atomic Energy Commission and other agencies of the Federal Government. Each issue contains two sections, the first is called "Technical Abstract Bulletin" and compiled by the Armed Services Technical Information Agency; the other section is called "Non-military and Older Military Research Reports" and lists new reports of the Atomic Energy Commission, the Office of Saline Water of the Department of the Interior and reports of other civilian agencies of the Government. In addition, this section lists military research reports not to be found in the "Technical

Abstracts Bulletin", most of these older military reports having been acquired in response to specific industry requests. As the title indicates no periodical literature is included. Most of the items include an abstract, in most cases the author's abstract is used. Each issue has a broad contents list using such headings as electronics and electronic equipment, fluid mechanics, nuclear physics and nuclear chemistry, physics, but there is no detailed subject index.

For abstracts concerned with specialized topics, see the appropriate chapter.

Questions

1. Why are abstracts important?
2. What different types of abstracts are there?
3. Compare the indexes of *Physics Abstracts* and *Nuclear Science Abstracts*.
4. Search *Physics Abstracts* for references to the rheology of milk gel.
5. Compare the coverage of *Nuclear Science Abstracts*, *Physics Abstracts*, *Engineering Index* and *Physikalische Berichte* for references to methods to improve the development characteristics of thick emulsions (400μ).

Societies, Research Organizations and Information Centres

Dewey 530.6 Class, etc.

SOME of the general publications of interest to the physicist have been described, but these are by no means the only sources of information. Information exists in two forms, the written form and unwritten "know-how". Fortunately, there are a number of published guides to sources of unpublished research—societies, research organizations and information centres.

Societies

Directories specifically covering Great Britain include:

Scientific and Learned Societies of Great Britain, Allen & Unwin, London, 1962. Grouping is under broad subject headings.
Trade Associations and Professional Bodies of the U.K., GB Research, London, 1962. This contains an alphabetical sequence of 1800 entries with a subject index enabling the societies in a particular field to be easily identified.

Societies in Great Britain for the physicist include the Royal Society, the Institute of Physics and the Physical Society, and the Royal Society of Edinburgh.

The Institute of Physics and the Physical Society is an amalgam of originally separate bodies which joined together in 1961. Its object is to promote the advancement of physics, to disseminate knowledge on physics and to elevate the profession of physics. It is the recognized British organization for consultations on all matters affecting the practice of the profession of physics. A person wishing to participate in its professional or educational work must belong to one of the professional grades of membership of the Institute of Physics. Meetings are organized at home and abroad. Specialist meetings are held by special subject groups for acoustics, applied spectroscopy, education, electron microscopy, electronics, low temperatures, non-destructive testing, optics, stress analysis and X-ray analysis.

The Royal Society was founded in 1660 and is the oldest society to have been in continuous existence. Its main object is the promotion of sciences. The Society grew from the periodical gatherings of men who believed that the correct way of investigating happenings was by observation and experiment. The first President was Lord Brouncker who has been followed by such eminent personages as Isaac Newton, under whose presidency the Society rose in numbers and prestige, Joseph Banks, Sir Humphry Davy, Rutherford, etc.

The Royal Society is responsible for the ultimate direction of the scientific work of the National Physical Laboratory, helps to manage the Royal Observatory, takes part in the work of the United Nations Educational, Scientific and Cultural Organization and the International Council of Scientific Unions, and is consulted by the Government on various scientific matters of national importance. Membership of the Society is only open to Fellows. There are about 600 Fellows and only 25 new ones can be admitted annually. The Society issues a year-book.

There are many books dealing with the Royal Society, including:

The Royal Society Tercentenary. Compiled from a special supplement of *The Times*, July 1960. Times Publishing, London, 1961.

ANDRADE, E. N. DA C., *A Brief History of the Royal Society*, Royal Society, London, 1960.

HARTLEY, SIR HAROLD (ed.), *The Royal Society, its Origins and Founders*, Royal Society, London, 1960.

LYONS, SIR HENRY, *The Royal Society, 1660–1940, a History of its*

Administration under its Charters, Cambridge University Press, Cambridge, 1940.

STIMSON, D., *Scientists and Amateurs: a History of the Royal Society,* Schuman, New York, 1948.

The Royal Society of Edinburgh was granted a charter by King George III in March 1783. The Fellows were originally divided into two classes—physical and literary—but as few literary papers were submitted separate meetings of this class soon ceased to be held. Clerk Maxwell, as a boy of 15, wrote his first paper for the Society on the "properties of certain oval curves". At present only twenty-five new Fellows can be admitted annually. A very readable account of the growth of the Society by J. Kendall is "The Royal Society of Edinburgh", *Endeavour,* **5**, 54 (1946).

Other societies and bodies of interest to the physicist include the British Astronomical Association, British Institute of Radiology, British Interplanetary Society, British Optical Association, Faraday Society, Institution of Electronics, Royal Astronomical Society, Royal Meteorological Society, Royal Photographic Society of Great Britain and the Royal Institution of Great Britain.

Directories covering the United States of America include:

Scientific and Technical Societies of the United States and Canada, National Research Council, Washington D.C., 7th ed., 1960–61.

Encyclopaedia of American Associations: a Guide to Trade, Professional, Labor, Scientific, Educational, Fraternal and Social Organizations, Gale Research Company, Detroit, 3rd ed., 1961.

BATES, R. S., *Scientific Societies in the United States,* Wiley, New York, 2nd ed., 1958.

The major physics body in the United States is the American Institute of Physics. This body was founded in 1931 by the Acoustical Society of America, American Association of Physics Teachers, American Physical Society, Optical Society of America and the Society of Rheology. In 1946, it was reorganized to provide membership for other organizations and individuals. Its object is the advancement and diffusion of knowledge of the science of physics and its application to human welfare.

Associate member societies of the American Institute of Physics are the American Crystallographic Association and the American Astronomical Society, and affiliated are the American Society for Metals, Cleveland Physics Society, Electron Microscope Society of America, Foundation for Instrumentation Education and Research, Geological Society of America, Philosophical Society of Washington, Physical Society of Pittsburgh, Physics Club of Chicago, Physics Club of Lehigh Valley, Physics Club of Philadelphia, Sigma Pi Sigma.

Guides to scientific societies exist for most countries, for example:

WANG, C., *Mainland China Organizations of Higher Learning in Science and Technology and their Publications: a Selected Guide*, U.S. Government Printing Office, Washington D.C., 1961.

WYLIE, G. J. and LOEW, N.F., *Australian Scientific Societies and Professional Organisations*, Commonwealth Scientific and Research Organization, Melbourne, 1951.

Scientific and Technical Associations and Institutes in Israel, National Council for Research and Development, Tel Aviv, 1962.

Scientific and Technical Societies in Japan, Maruzen, Tokyo, 1962.

Scientific Institutions and Scientists in Latin America, Centro de Cooperacion Cientifica pera America Latina, Montevideo, 1947–.

International Organizations

Among international organizations of value to the physicist are:

International Astronomical Union. This body was founded in 1919 and aims to facilitate relations amongst astronomers in cases where international co-operation is useful and to promote the study of astronomy in all its branches. Over twenty-five countries are members, the United Kingdom being represented by the Royal Society and the United States of America by the National Research Council. The Union has commissions made up of various members who investigate topics of particular interest, e.g. astronomical constants, solar eclipses, physical observations of comets, stellar spectra, radioastronomy, history of astronomy, etc.

International Council of Scientific Unions. Founded in 1919 as the International Research Council, it assumed its present title in 1931. Through its members, it aims to direct international scientific activity in subjects which do not fall within the purview of any existing international associations and to enter in relation with the governments of the countries adhering to the Council in order to promote scientific investigation in these countries. The Council has under it joint Commissions in borderline fields between two or more unions such as ionosphere, radiometeorology, solar and terrestrial relationships and applied radioactivity. The Council also institutes projects such as the International Geophysical Year.

International Foundation of the High Altitude Research Station Jungfraujoch. In 1930 the foundation was created and the research station opened in 1931. Its aims are to promote all fields of science interested in high altitude experimentation, e.g. astrophysics, geophysics, physics, etc.

International Union of Geodesy and Geophysics. The Union is a confederation of seven International Associations: International Association of Seismology, International Association of Meteorology, International Association of Geodesy, International Association of Terrestrial Magnetism and Electricity, International Association of Physical Oceanography, International Association of Vulcanology, International Association of Hydrology. It was formed in 1919 to promote the study of problems relating to the figure and physics of the earth, to initiate, facilitate and co-ordinate research into and investigation of those problems of geodesy and geophysics which require international co-operation. The primary activities are carried out by the constituent International

Associations and Commissions in their various fields of scientific investigation.

International Union of Pure and Applied Physics. The Constitutive Assembly of the Union was held in 1923 and the objects of the Union are to create and encourage international co-operation in physics, to co-ordinate the work of preparation and publication of abstracts of papers and of tables of physical constants, to bring about international agreement on matters of units, standards, nomenclature and notations and to support research in suitable directions. Commissions consider such topics as optics, thermodynamics and statistical mechanics, symbols, units, etc.

United Nations Educational Scientific and Cultural Organization. Came into being in 1946 and has governmental representatives as its members. Amongst the work of the Organization is the facilitation of scientific work and particularly research, as well as the dissemination and exchange of scientific information between each particular region and the rest of the world, and this is done through the Field Service Co-operation Offices—the work of these Offices can be followed in *Science Liaison: The Story of UNESCO's Science Co-operation Offices*, UNESCO, Paris, 1954. UNESCO makes grants-in-aid to international scientific organizations, is interested in scientific documentation, the dissemination and teaching of science and has an Exchange of Persons Service which helps to promote international understanding through the exchange of persons between one country and another for education purposes.

Guides detailing the activities of international organizations include:

Directory of International Scientific Organisations, UNESCO, Paris, 2nd ed., 1953.

Guide to International Organisations, Central Office of Information, London.

Yearbook of International Organisations, Union of International Organizations, Brussels.

Governmental Research

In Britain, one of the best sources of information on physics and related topics is the National Physical Laboratory. This is a station of the Department of Scientific and Industrial Research and its terms of reference are "to carry out research in those areas which are of national importance, but which for various reasons, may be less effectively studied by other less centrally placed organisations". The annual report details the work being done by the various divisions: aerodynamics, applied physics, autonomics, basic physics, light, mathematics, metallurgy, ships, and standards.

Another famous national laboratory is the National Institute for Research in Nuclear Science whose current work is outlined in their annual report, whilst information on other government laboratories may be found from *Industrial Research in Britain*, Harrap, London, 4th ed., 1962.

Covering governmental research work in the United States is *Federal Organisation for Scientific Activities 1962*, U.S. Government Printing Office, Washington D.C., 1963. This reports on the organization and programme content for science of the 40 Federal Agencies involved in scientific activities and covers research and developments, scientific and technical information, research data collection, and scientific testing and standardization.

University and Technical College Research

To get a lead on to particular universities or technical colleges in Britain who are doing work of interest one would consult the yearly publication *Scientific Research in British Universities*, HMSO, London. Under the general heading of a particular university or technical college the various departments are listed with the work being undertaken in each department and the name of the scientist responsible for the work. A name and subject index is included so that it is easy to find if work is being undertaken on a particular subject. The similar publications covering the U.S.A. include *Graduate Physics Research Specialities in*

American Educational Institutions, American Institute of Physics, New York (yearly) and *Directory of University Research Bureaux and Institutes*, Gale Research, Detroit.

Other guides include such as *Schweizerischer Hochschulkalender*, Leeman, Zürich, which covers Swiss universities and their associated research institutes.

Industrial Research

To locate research work undertaken by British industry the best source is *Industrial Research in Britain*, Harrap, London, 4th ed., 1962, one section of which gives information about the research activities of some industrial firms. A subject index increases the value of this publication. The United States is covered by *Industrial Research Laboratories of the United States*, National Research Council, Washington D.C., 1960. All the firms having research facilities are not listed in these two publications. To obtain the names of firms who might have information on a particular topic, the various trade directories should be consulted, such as *British Instruments Directory and Buyers Guide*, United Science Press, London, and *Communications and Electronics Buyers Guide and Who's Who*, Heywood, London. Another source of industrial contacts would be *Aslib Directory*, Aslib, London, which contains details of the libraries and subject fields of their industrial members, or in the case of the United States, *Directory of Special Libraries*, Special Libraries Association, New York, which gives the same details.

Technical Services for Industry, DSIR, London, 1962, gives a general survey of the work of the grant-aided research associations of which there are over fifty with interests ranging as wide as food, steel, jute, printing, paint and scientific instruments. A subject index is included leading easily from a subject to the body working in that field. *Research for Industry*, HMSO, London, is another useful guide.

The British Scientific Instrument Research Association in conjunction with the Scientific Instrument Manufacturers' Association operate an enquiry service available to all, of the "where to buy" nature on instruments, instrument components, laboratory apparatus and automatic control equipment of British origin.

Not all associations doing research in Britain are grant aided, e.g. National Rubber Producers Research Association, Dyers and Cleaners Research Organization. Some of these can be very valuable sources of information, though finding details of them is not easy, their names being included in some trade directories but not in others.

Trade Associations

Trade associations in the United States are listed in the *Encyclopaedia of Associations: Volume 1. National Organisations of the U.S.*, Gale Research, Detroit, 3rd ed., 1961, whilst for Great Britain one consults *Trade Associations and Professional Bodies of Great Britain*, GB Research, London, 1962. A trade association will sometimes be able to supply information directly or will circularize its members asking for help on the particular problem.

Directories

Bibliography of Directories of Sources of Information, International Federation for Documentation, The Hague, 1960.

Guide des Centres Nationaux d'Information Bibliographique, UNESCO, Paris, 1953.

Guide to European Sources of Information, OECD, Paris, 1957, which lists centres which are prepared to furnish information on technological, industrial and management questions. There is an index by subject and the scope of each centre is described, including its research.

Handbook of Current Research Projects in the Republic of China, National Central Library, Taipei, 1962.

Index Generalis, Dunod, Paris, *Minerva*, de Gruyter, Berlin, and *World of Learning*, Europa Publications, London, cover universities, learned societies, etc., in many countries.

Information Booklet on Physics Organisations Abroad, American
Institute of Physics, New York, 1962, details the work of
laboratories, societies and institutions.

Instituti e Laboratori Italiani di Ricerca Scientifica, National Research Council, Rome, 1953.

Jaarboek van de Nederlandse Nataurkundige Vereniging, Nederlandse Natuurkundige Vereniging, Amsterdam, carries information on physics societies, laboratories and research organizations of the Netherlands.

Répertoire de Laboratories Scientifiques, L'Éducation Nationale,
Paris, 1962.

Research Centres, Commonwealth Scientific and Industrial Research Organizations, Melbourne, 1953. This lists 295 organizations, including CSIRO establishments, research and industrial associations, private research organizations, departments
of state governments and public authorities, universities, etc.

Scandinavian Research Guide, Scandinavian Council for Applied
Research, Oslo, 1960, gives details of the activities of various
research institutes, laboratories, central research organizations,
universities, institutes of technology, information centres and
scientific societies.

Science Information in Japan, Japan Documentation Society,
Tokyo, 1962. Sources such as societies and universities are
included.

Specialised Science Information Services in the United States,
U.S. Government Printing Office, Washington D.C., 1961.

Taschenbuch für das wissenschaftliche Leben, Festland, Bonn,
contains details of German scientific societies, universities and
other research institutes, and official organizations concerned
with science.

The Scientific and Academic World, Stifterverband für die Deutsche
Wissenschaft, Essen, 1962, lists universities and other institutions of higher learning, and industrial, private and governmental laboratories in over twenty countries.

Vademecum Deutscher Forschungsstätten, German Foundation
for Scientific Research, Essen, 1954.

Details are given of the area of specialization, types of information services provided and whether there is any restriction on the availability of information.

Information Centres

National information centres include:

The National Referral Center established in the Library of Congress, Washington D.C., to advise an enquirer where to go for data in a scientific field or on a technical subject of interest to him. For information on the current federally supported research, one could approach the Office of Science Information Service, National Science Foundation, Washington D.C.

The Department of Scientific and Industrial Research, Headquarters Information Division, State House, High Holborn, London, W.C. 1, provides a technical enquiry service to direct enquirers to appropriate literature or specialist bodies that can help.

Technical information centres have also been established in the following countries:

AUSTRIA
OPZ, Wien, 1, Renngasse 5.

BELGIUM

OBAP, 60 rue de la Concorde, Bruxelles.

CANADA
Technical Information Service, National Research Council, Ottawa 2, Ontario.

DENMARK
DTO, Ørnevej 30, Copenhagen NV.

FRANCE
AFAP, Renseignements Techniques et Industriels, 11 rue du Fg. Saint-Honoré, Paris 8ème.

GERMANY
RKW, Frankfurt a/m. Gutleutstrasse 163–167.

GREECE
Centre Hellénique de Productivité, 28 rue Kapodistriou, Athènes.

ICELAND
Idnadarmalastofnun Islande, Post Box 160, Reykjavik.

IRELAND
Institute for Industrial Research and Standards, Glasnevin House, Ballymun Road, Dublin.

ITALY
CNP, Casella Postale 2472 AD, Rome.

LUXEMBURG
OLAP, 60 rue Auguste-Lumière, Luxemburg.

NETHERLANDS
NTS, 6 Willem Witsenplein, La Haye.

NORWAY
SNI, Forskningsveien 1, Blindern-Oslo.

PORTUGAL
Comisao Técnica de Cooperacao Economica e Externa, Avenide da Republica 32–1°, Lisbon.

SPAIN
CID, Patronato "Juan de la Cierva", Madrid 6–Serrano 150.

SWEDEN
Ingeniörsvetenskapsakedemien, SIU, Box 5073, Stockholm 5.

SWITZERLAND
Office Suisse d'Expansion Commerciale, Dreiköningstr 8, Zürich.

TURKEY
Türk Teknik Haberlesme Merkesi, Istanbul Technical University, Istanbul.

YUGOSLAVIA

Head of Foreign Relations Section, Yugoslav Productivity Institute, Uzun Kirkova 1, Belgrade.

For societies, research associations and information centres concerned with specialized topics, see the appropriate chapter.

Questions

1. Why are sources other than the literature of value when seeking information?
2. What research work in physics is being undertaken at Manchester University?
3. If information is required on industrial research laboratories in the United States of America, where is such information available?
4. Detail Japanese sources of scientific information.
5. What purposes do societies fulfil?
6. Which Italian societies are of interest to the physicist?

CHAPTER 8

Relativity, Quantum Mechanics, Statistical Mechanics, Mathematical Physics, Experimental Design and Instruments

Dewey 530.11, 530.12, 530.13, 530.15, 530.78 Classes

Relativity [Dewey 530.11 Class]

The theory of relativity is largely due to Albert Einstein; it is a theory of the physical meaning of space and time and has two parts: (1) the special theory, which explains why the laws of nature appear the same to all observers moving with a constant velocity to one another, (2) the general theory, which is the relativistic theory of gravitation and an extension of the special theory. The special theory has practical use in modern physics whilst the general theory has led to the study of cosmology—the nature and origin of the universe.

For information one cannot do better than consult the works of the originator of the theory—Einstein's *Relativity*, Methuen, London, 15th ed., 1955, and *The Meaning of Relativity*, Methuen, London, 6th ed., 1956. Should a layman's guide be required, C. V. Durrell's *Readable Relativity*, Bell, London, 1962, has been called the best layman's guide ever written, and others of an introductory nature are:

BARNETT, L., *The Universe and Dr. Einstein*, Gollancz, London, 6th ed., 1959.

LANDAU, L. D. and RUMER, G. B., *What is Relativity?* Oliver & Boyd, London, 1960.

There are many books on relativity and amongst those suitable for the serious student are:

SYNGE, J. L., *Relativity: the Special Theory*, North-Holland, Amsterdam, 1956.

SYNGE, J. L., *Relativity: the General Theory*, North-Holland, Amsterdam, 1960.

The early literature can be traced from the bibliography *Bibliographie de la Relativité*, Lamertin, Bruxelles, 1924, whilst current thinking is surveyed in *Recent Developments in General Relativity*, Pergamon, London, 1962. The latter publication contains reviews of such problems as gravitational radiation, problems of motion, experimental tests, quantization, unified theories, cosmology and mathematical problems of general relativity, and also original contributions surveying the different schools of thought, different mathematical methods and scientific approaches to various problems of general relativity.

Quantum Mechanics [Dewey 530.12 Class]

Quantum mechanics uses the common features of wave mechanics and matrix mechanics in a system which can be used to produce different forms of theory which may be useful in special cases. De Broglie initiated quantum mechanics in 1923 and it was developed by Dirac in 1925, followed in 1926 by Heisenberg's matrix mechanics and Schrodinger's wave mechanics, which in turn were followed by the physical interpretation of the waves.

One of the most prolific writers on this subject is Louis de Broglie and any of his works can be recommended but especially his *Physics and Microphysics*, Hutchinson, London, 1955, which outlines the developments of wave mechanics. The writings of the developers of the theory can be studied from texts such as:

DIRAC, P. A. M., *The Principles of Quantum Mechanics*, Oxford University Press, London, 4th ed., 1958.

HEISENBERG, W., *The Physical Principles of the Quantum Theory*, University of Chicago Press, Chicago, 1930.

SCHRODINGER, E., *Collected Papers on Wave Mechanics*, Blackie, London, 1928.

For the undergraduate, C. W. Sherwin's *Introduction to Quantum Mechanics*, Winston, New York, 1959, and R. M. Sillito's *Non-Relativistic Quantum Mechanics*, Edinburgh University Press, Edinburgh, 1960, are valuable, whilst the research worker would find any of the following of value:

BATES, D. R. (ed.), *Quantum Theory*, Academic Press, New York, 1961.

HEINE, V., *Group Theory in Quantum Mechanics: an Introduction to its Present Usage*, Pergamon, London, 1960.

Statistical Mechanics [Dewey 530.13 Class]

Statistical mechanics endeavours to explain the macroscopic properties of a system on the basis of the properties of the microscopic constituents of the system. It differs from kinetic theory in that it uses a more sophisticated picture of the atom and refined techniques and results from the mathematical theory of probability.

J. Willard Gibbs had considerable influence on modern statistical mechanics and his works should be consulted by any serious student of the subject. This can be done fairly easily since they have all been brought together in *The Collected Works of J. Willard Gibbs*, Longmans, New York, 1928. More up-to-date views are given in a series of studies edited by J. de Boer and G. E. Uhlenbeck which reflect recent progress and the search for basic understanding, *Studies in Statistical Mechanics*, North-Holland, Amsterdam, 1962.

From the many books on statistical mechanics the following are an indicative selection but it is certainly not meant to imply that those not mentioned have no value:

CHISHOLM, J. S. R. and BORDE, A. H. DE, *An Introduction to Statistical Mechanics*, Pergamon, London, 1958.

FOWLER, R. H., *Statistical Mechanics: the Theory of the Properties of Matter in Equilibrium*, Cambridge University Press, London, 3rd ed., 1936.

HILL, T. L., *Statistical Mechanics: Principles and Selected Applications*, McGraw-Hill, New York, 1956.

KITTEL, C., *Elementary Statistical Physics*, Wiley, New York, 1958.

PRIGOGINE, I., *Non-Equilibrium Statistical Mechanics*, Interscience, New York, 1961.

RUSHBROOKE, G. S., *Introduction to Statistical Mechanics*, Oxford University Press, London, 1949.

SCHRODINGER, E., *Statistical Thermodynamics*, Cambridge University Press, London, 2nd ed., 1952.

Mathematical Physics [Dewey 530.15 Class]

Mathematical physics is concerned with describing physical phenomena in mathematical language and was a field in which Wolfgang Pauli was particularly interested. M. Fierz and V. F. Weisskopf have edited a memorial volume to him; *Theoretical Physics in the Twentieth Century*, Interscience, New York, 1960. Some of the articles summarize the progress in some of the topics in which Pauli was interested and some are reports of the "heroic" period of physics during the 1930s.

Other texts which can be recommended include:

BAND, W., *Introduction to Mathematical Physics*, Van Nostrand, Princeton, New Jersey, 1959.

HOUSTON, R. A., *An Introduction to Mathematical Physics*, Blackie, London, 1952.

KOMPANEYETS, A. S., *Theoretical Physics*, Foreign Languages Publishing House, Moscow, 1961.

LINDSAY, R. B., *Concepts and Methods of Theoretical Physics*, Van Nostrand, Princeton, New Jersey, 1951.

RICHARDS, P. I., *Manual of Mathematical Physics*, Pergamon, London, 1959.

SLATER, J. C. and FRANK, N. H., *Introduction to Theoretical Physics*, McGraw-Hill, New York, 1933. Revised editions have been issued in four volumes: mechanics, electromagnetism, quantum theory and chemical physics.

Methods of doing calculations in physics are the subject of several books, but it will be sufficient here to mention merely two as an indication of the sort of help which is available. In their *Calculations in Physics*, Murray, London, 1959, W. Ashurst and H. A. Hornby have taken problems set for the Advanced Level of the General Certificate of Education and V. L. Zubov and V. P. Shal'nov focus attention on those points which give rise to most errors in examples taken from Russian secondary schools and Moscow University examinations in their *Worked Examples in Physics*, Pergamon, London, 1962.

Periodicals of specific interest include *Journal of Research: Section B, Mathematics and Mathematical Physics*, which is issued quarterly by the National Bureau of Standards and contains studies and compilations designed mainly for the mathematician and theoretical physicist, and each issue contains abstracts of the Bureau's publications. The *Journal of Mathematical Physics* is issued monthly and shows the advances in mathematical techniques applicable to the various branches of modern physics.

Experimental Design and Instruments
[Dewey 530.78 Class]

It is very important when conducting experiments in physics, or indeed in any science, that a reasonable period beforehand be spent on a definite analysis of the purpose of the experiment, the possible methods and the number and accuracy of the measurements to be made. Otherwise it may be found after considerable time has been spent that the apparatus and methods are in fact not capable of producing the required results.

H. J. J. Braddick outlines the principles of experimentation in his *The Physics of Experimental Method*, Chapman & Hall,

London, 1954, and he lays emphasis on the physical principles of measurement, the reduction of observations and on the statistical analysis of errors. The essential dependence on the properties of the various materials used in the construction of apparatus is also brought out.

The statistical view of experimental design is covered both by D. R. Cox in *Planning of Experiments*, Wiley, New York, 1958, and B. J. Winer in *Statistical Principles of Experimental Design*, McGraw-Hill, New York, 1962.

The design of instruments and apparatus, and their uses, was the subject of the 5th International Conference held in Stockholm in September 1960, and the proceedings have been published, under the editorship of H. Von Koch and G. Ljungberg, as *Instruments and Measurements*, Academic Press, New York, 1961. The same field is also covered by H. J. J. Braddick in *Mechanical Design of Laboratory Equipment*, Chapman & Hall, London, 1960, and another from the same publisher by A. Elliot and J. Home Jackson, *Laboratory Instruments: their Design and Application*, Chapman & Hall, London, 2nd ed., 1959.

Periodicals which include information on instrument design and use include the *Review of Scientific Instruments*, which is issued by the American Institute of Physics every month; the *Journal of Scientific Instruments*, which includes original work, reviews, historical notes, technical news, book reviews and correspondence on the principles, construction and use of scientific instruments, and is issued monthly by the Institute of Physics and the Physical Society in association with the National Physical Laboratory; the *Transactions of the Society of Instrument Technology;* the *Instrument Society of America Journal; Instrument Practice; Instrument Engineer;* and the cover-to-cover translation of the Russian periodical *Priborostroenie* published in English as *Instrument Construction.*

The British Scientific Instrument Research Association studies fundamental techniques, evaluates instruments and materials, and develops and constructs novel and special instruments. *Instrument Abstracts* is compiled by the Association and published monthly by Taylor & Francis, London. Abstracts are included covering astronomy, atomics and nucleonics, biology and medicine, chemistry, control, data handling, electricity and electronics, fluid mechanics, geophysics, and meteorology, heat, light, magnetism and electromagnetism, materials and design, mechanics, navigation, surveying and telecommunications. Annual author and subject indexes are issued.

Bibliographies which can usefully be consulted include:

BROMBACKER, W. G., *et al. Guide to Instrumentation Literature*, U.S. Government Printing Office, Washington D.C., 1955.

SMITH, J. F., *Instrument Literature and its Use: a Guide and Source List*. National Bureau of Standards, Washington D.C., 1953.

Methods of physical measurements are well covered in a six-volume series edited by L. Marton, *Methods of Experimental Physics*, Academic Press, New York, 1959. The volumes cover classical methods, electronic methods, molecular physics, atomic and electronic physics, nuclear physics and solid state physics. Another series of very useful books, being experiments selected from the *School Science Review* and suitable for beginners and sixth-form students, with particular parts covering physics, are published with the general title *The Science Master's Handbook*, Murray, London. Other books on practical physics include:

DAISH, C. B. and FENDER, D. H., *Experimental Physics*, English Universities Press, London, 1956.

YARWOOD, T. M., *Intermediate Practical Physics*, Macmillan, London, 3rd ed., 1958.

Questions

1. How does statistical mechanics help the physicist?
2. What aspects should be considered in the design of experimental apparatus?
3. What place has relativity played in modern physics?

CHAPTER 9

Mechanics, Sound
Dewey 531/533, 534 Classes

Mechanics [Dewey 531/533 Classes]

Mechanics is the science which deals with the effects of forces on bodies at rest or in motion. It is usually divided into the study of liquids, the study of the action of gases, and the study of rigid or elastic bodies. Frequently, the term "mechanics" is restricted to this latter field.

The latest developments can be followed from review publications such as *Advances in Applied Mechanics*, Academic Press, New York, and *Progress in Solid Mechanics*, North-Holland, Amsterdam.

For the layman seeking to understand something of mechanics, there are many general works on physics which include mechanics, such as *Physics*, Heath, Boston, 1960, or P. Abbot's *Teach Yourself Mechanics*, English Universities Press, London, 1959.

One of the classic texts on the subject is F. Auerbach's *Handbuch der physikalischen und technischen Mechanik*, Barth, Leipzig, 1927–31, but any of the following will be useful to students:

BULLEN, K. E., *Theory of Mechanics*, Science Press, Sydney, 4th ed., 1958.

ELEN, L. W. F. and MYERS, R., *Mechanics: a New Introduction*, Cleaver Hume, London, 2nd ed., 1958.

FOWLES, G. R., *Analytical Mechanics*, Holt, New York, 1962.

HUMPHREY, D. and TOPPING, J., *A Shorter Intermediate Mechanics*, Longmans, London, 2nd ed., 1961.

LANDAU, L. D. and LIFSHITZ, E. M. *Mechanics*, Pergamon, London, 1960.

LEVINSON, I. J., *Introduction to Mechanics*, Prentice-Hall, Englewood Cliffs, New Jersey, 1961.

The American Society of Mechanical Engineers publishes the quarterly *Journal of Applied Mechanics* and the bi-monthly *Journal of the Mechanics and Physics of Solids*, which deals with the theoretical principles governing the use of constructional materials and also experimental results.

Periodicals such as these are abstracted by the major abstracting services such as *Physics Abstracts* (Science Abstracts A) and also by the monthly specialized Russian abstracting services, *Referativnyi Zhurnal: Mekhanika*. *Applied Mechanics Reviews* carries reviews, articles and also abstracts grouped under subject headings such as elasticity, vibration of solids and incompressible flow. It is issued monthly by the American Society of Mechanical Engineers and has a yearly subject index. The American Society of Mechnical Engineers, which has a special division covering applied mechanics, the British Institution of Mechanical Engineers and other similar national bodies co-operate in the International Union of Theoretical and Applied Mechanics (IUTAM), whose functions are to act as a link between persons and national or international organizations engaged in work on mechanics, to organize international meetings and to promote the development of the science of mechanics. Members of the Union include the U.S. National Committee on Theoretical and Applied Mechanics, the Swedish National Committee for Mechanics, the Israel Society for Theoretical and Applied Mechanics and the National Committee for Theoretical and Applied Mechanics of the Science Council of Japan.

Research work is undertaken by Universities on such topics as celestial, fluid and statistical mechanics, and by laboratories such as the American National Bureau of Standards which works on engineering, fluid, pressure, vacuum and rheological aspects of mechanics.

SCIENCE (Specific Subjects)

* † ‡ UNIVERSITY OF CAMBRIDGE : CLARE COLLEGE. **Denman Baynes Research Studentship**, see [483].

* DEPARTMENT OF SCIENTIFIC AND INDUSTRIAL RESEARCH. **Grants**, see [124].

* † ‡ UNIVERSITY OF DURHAM : KING'S COLLEGE, NEWCASTLE UPON TYNE. (i) **Associated Lead Manufacturers Limited Student ship**, see [486].

(ii) **George Angus Studentship**, see [125].

* † ‡ UNIVERSITY OF HULL. **British Coke Research Association Fellowship** and **University Research Studentships**, see [488]-[489].

* † ‡ IMPERIAL CHEMICAL INDUSTRIES, LTD. **I.C.I. Research Fellowships**, see [490].

* UNIVERSITY OF KEELE. **Research Studentships**, see [37].

* † ‡ UNIVERSITY OF LEEDS. **Henry Ellison Senior Fellowship**, see [525].

* † ‡ UNIVERSITY OF LONDON INSTITUTE OF CANCER RESEARCH : ROYAL CANCER HOSPITAL. **Research Studentships**, see [437].

* † ‡ UNIVERSITY OF LONDON : QUEEN ELIZABETH COLLEGE. **Postgraduate Research Scholarships**, see [493].

[553] UNIVERSITY OF LONDON : QUEEN MARY COLLEGE

* † ‡ **British Oxygen Research Studentship**

Subject. Low-temperature physics.

Value. About £335 p.a.

Tenable at Queen Mary College, for not more than 3 years normally.

Eligibility. Open only to physics graduates. Candidates need not be in U.K. in order to be eligible.

Number. 1 offered annually for 7 years from and including 1959–60.

Closing date. 30th June.

Particulars from Registrar, Queen Mary College, Mile End Road, London, E.1.

[554] UNIVERSITY OF LONDON : ROYAL FREE HOSPITAL SCHOOL OF MEDICINE

* † ‡ **William Fletcher Barrett Research Scholarship**

Subject. Physics.

Value. £60 p.a.

Tenable in Department of Medical Physics, Royal Free Hospital School of Medicine, for 2 years ; may be extended for third year.

Eligibility. Candidates need not be in U.K. in order to be eligible.

Number. 1 offered for 1962–63, 1 (if no award in 1962) for 1963–64. Offered every 2 years if award not extended.

Closing date. 1st May.

Particulars from Secretary, Royal Free Hospital School of Medicine, Hunter Street, London, W.C.1.

† UNIVERSITY OF MELBOURNE. **Sir Arthur Sims Travelling Scholarship**, see [62].

† NATIONAL RESEARCH COUNCIL OF CANADA. **Postdoctorate Overseas Fellowships** and **Special Scholarships**, see [494]-[495].

* † ‡ UNIVERSITY OF NOTTINGHAM. **Captain Black Science Research Scholarship**, see [496].

* † ‡ UNIVERSITY OF OXFORD. **Pressed Steel Research Fellowships**, see [498].

[555] UNIVERSITY OF READING

* † ‡ **Associated Electrical Industries Research Fellowship or Scholarship**

Subject. Physics.

Value. Not yet determined.

Tenable at University of Reading for 3 years normally, but can be held for shorter period.

Eligibility. Candidates need not be in U.K. in order to be eligible.

Number not specified. Offered as vacancy occurs.

Particulars from Professor of Physics, J. J. Thomson Physical Laboratory, University of Reading, Whiteknights Park, Reading.

* † ROYAL SOCIETY. **Rutherford Scholarship**, see [694].

* UNIVERSITY OF SHEFFIELD. **Charles Kingston Everitt Memorial Scholarship**, see [132].

* † ‡ UNIVERSITY OF SOUTHAMPTON. **University Postgraduate Research Awards**, see [88].

* † ‡ TURNER AND NEWALL, LTD. **Research Fellowships**, see [503].

* † ‡ UNIVERSITY OF WALES : UNIVERSITY COLLEGE OF WALES, ABERYSTWYTH. **Sir Garrod Thomas Fellowship**, see [532].

FIG. 1. *United Kingdom Postgraduate Awards, 1962–64.*

Physics

1 Summary of Work

Physics deals with the properties of matter and energy. As a school subject its main divisions are mechanics, heat, light, sound, magnetism and electricity: these headings, together with electronics and atomic or nuclear physics, indicate the main fields of employment. Physics provides the theoretical basis of engineering science and physical laws apply in all branches of natural science. The field of the physicists' work is therefore very broad and the mathematical aspects of it are often important. Physicists' work ranges from practical experimentation to what is hard to distinguish from mathematics.

2. Qualities and Educational Qualifications Required

Patience, perseverance and the ability to concentrate are valuable as they are in any responsible work; practical skill in experimental work is an asset. Powers of observation, imagination and reasoning power are also required. A sound training in scientific method is important. The measurement of physical properties and relationships calls for accuracy and precision.

In order to become a qualified professional physicist it is essential to have the ability to undertake studies to graduate or professional level. All intending physicists are advised to remain at school to take Advanced level or (in Scotland) Higher grade sciences (physics, pure mathematics, applied mathematics and preferably chemistry) in order to secure admission to degree or comparable courses of study.

3. Training

Training to graduate standard is best done by full-time study. Most students work for a university degree. All universities offer degree courses in physics. (At Scottish universities the appropriate course may be in natural philosophy and mathematics.) An equivalent qualification is the Diploma in Technology and an increasing number of students are taking this by full-time or sandwich courses at technical colleges. In addition many technical colleges offer full-time or sandwich courses leading to the graduateship examination of the Institute of Physics and The Physical Society or the external degree of London University.

Candidates who wish to become professional members of The Institute of Physics and The Physical Society (Associates—A.Inst.P.) must:

(1) have attained age 25 years;

(2) have had experience (normally three-five years) in responsible work in physics or its applications; and

(3) possess suitable academic qualifications.

Fig. 2. *Careers Guide.*

DENSITY AND THERMAL EXPANSION OF CHEMICAL COMPOUNDS IN THE CRYSTALLINE STATE, UNDER ATMOSPHERIC PRESSURE

J. R. CLARKE

I. Inorganic Compounds (C-Table)

This section includes only substances for which density values reliable to four decimal places, or values over a temperature range, are available. For less accurate values at individual temperatures, v. Vol. I, p. 106.

Symbols and Abbreviations

The quantities recorded in the table below are the following:

$$\alpha^t = \frac{10^4}{l}\frac{dl}{dt} \text{ at } t^\circ; \text{ or } \alpha = \frac{10^4\Delta l}{l\Delta t} \text{ over the range } \Delta t^\circ.$$

$\Delta = 10^9 \frac{d\alpha}{dt}\left(\text{resp. } 10^9 \frac{\Delta\alpha}{\Delta t}\right)$, $i.e.$, $10^9 \times$ the rate of change of α with t, at t° (resp. over Δt°).

d_4^t = the density in g/ml at t° = the specific gravity at t° referred to H_2O at 4°C.

d_o, a, b and c_i = the parameters of the equation: $d_4^t = d_4^0(1 - 10^{-5}$ $at - 10^{-7} bt^2 - 10^{-9}ct^2)$, valid over the range indicated; b and c are zero when not given.

Arrangement

The compounds are arranged in groups, as follows: I. Oxides; II. Halogen compounds in the order, F, Cl, Br, I; III. Sulfides; IV. Sulfates; V. Nitrates; VI. Carbonates; VII. Metallo-organic compounds; VIII. Various silicates; IX. Other compounds. For organic compounds, v. the C-Table, p. 45.

I. OXIDES

H_2O	$d_4^0 = 0.9168 \pm 0.0005;^*$ For α see Fig. 1
H_2O_2	$d_4^0 = 1.6436$ [34]
As_2O_3	$\begin{cases} d_4^0 = 3.873, a = 12 \ (0^\circ-50^\circ) \ [3] \\ \alpha^{40} = 41.26, \Delta(20^\circ-70^\circ) = 67.9 \ [17, 18, 33] \end{cases}$
Sb_2O_3: Senarmontite	$\alpha^{40} = 19.6s, \Delta(20^\circ-70^\circ) = 5.7 \ [17, 18, 33]$
CO_2	$\begin{cases} t, {}^\circ C \mid -80 \mid -90 \mid -100 \mid -110 \mid -120 \mid -130 \mid -183 \\ d_4^t \mid 1.565 \mid 1.581 \mid 1.594 \mid 1.607 \mid 1.618 \mid 1.627 \mid 1.66s \end{cases}$ [35.5]
SiO_2	v. Vol. IV, p. 19.
TiO_2: Brookite	Values of $\alpha^{17.5}$ ∥ a, b and c axes (a) 14.4939, (b) 19.2029, (c) 22.0489 [44]
Anatase	$\alpha^{40} = 8.19 \parallel, = 4.68 \perp;$ and $\Delta(20^\circ-70^\circ) = 31.1 \parallel, = 29.5 \perp$ to opt. axis [33]; $cf.$ [44]
Rutile	$\alpha^{40} = 7.14 \perp, = 9.19 \parallel; \Delta(20^\circ-70^\circ) = 11.0 \perp, = 22.5 \parallel$ to principal axis [17, 18, 33]; $cf.$ [44]
ZrO_2	[35.5]
SnO_2: Cassiterite	$\alpha^{40} = 3.2 \perp, = 3.9 \parallel; \Delta(20^\circ-70^\circ) = 7.6 \perp, = 12 \parallel$ to principal axis [17, 18, 33]
PbO_2	$\alpha(25^\circ-93^\circ) = 7.9 \pm 0.6 \ [39]$.
ThO_2	[36.5]
ZnO: Zincite	$\alpha^{40} = 3.2 \perp, = 3.9 \parallel; \Delta(20^\circ-70^\circ) = 7.6 \perp, = 12 \parallel$ to principal axis [17, 18, 33]
Cu_2O: Cuprite	$\alpha^{40} = 0.93, \Delta(20^\circ-70^\circ) = 21. \ d_{max.}$ at $-43^\circ C$ [15, 16, 17, 18, 21, 22, 33]; $cf.$ [42]
Fe_2O_3: Hematite	$\alpha^{40} = 7.61 \parallel = 7.71 \perp; \Delta(5^\circ-80^\circ) = 4.90 \parallel, = 11.4 \perp$ to opt. axis. [3]; $cf.$ [17, 18, 33]
Fe_3O_4: Magnetite (Fe, Zn, Mn) O.Fe$_2$O$_3$: Franklinite	$\alpha^{40} = 8.46, \Delta(20^\circ-70^\circ) = 28.9 \ [14, 33]$ $\alpha^{40} = 8.1, \Delta(20^\circ-70^\circ) = 9.4 \ [14, 33]$
Al_2O_3: Corundum	$\alpha^{40} = 6.2 \parallel; = 5.4 \perp;$ and $\Delta(20^\circ-70^\circ) = 20.5 \parallel; = 22.5 \perp$ to principal axis. [14, 15, 16, 33]; $cf.$ (Vol. II, p. 87)
$Al_2O_3.ZnO$: Gahnite 3Al$_2$O$_3$.2FeO.4SiO$_2$: H_2O: Staurolite	$\alpha^{40} = 5.95, \Delta(20^\circ-70^\circ) = 18.3 \ [14, 15, 16, 33]$ $\alpha^{40} = 7.0s, \Delta = 31.s \ [14]$

*Estimate by Howard T. Barnes on basis of critical examination of all available data. For discussion and bibliography see [3.5] and his forthcoming book on *The Physics of Ice.*

Right column top table:

(Zn, Fe)O.(Al, Fe)$_2$O$_3$: Kreittonite		$\alpha^{40} = 5.96, \Delta(20^\circ-70^\circ) = 19.4$ [15, 16, 33]			
$Al_2O_3.BeO$: Chrysoberyl	∥	α'-axis	α''-axis	α'''-axis†	Lit.
	α^{40}	5.16	6.01	6.02	[14]
	Δ	12.2	10.1	22.0	[33]
MgO	(Prefused) $\alpha = 11.40 + 9.2(t - 120), 120^\circ$ to 270° [24]				
Periclase	$\alpha^{40} = 10.43, \Delta(20^\circ-70^\circ) = 26.7$ [15, 16, 33]				
$MgO.Al_2O_3$: Spinel (Mg, Fe)O.Al$_2$O$_3$: Pleonaste	$\alpha^{40} = 5.93, \Delta(20^\circ-70^\circ) = 19.5$ [15, 16, 33] $\alpha^{40} = 6.03, \Delta(20^\circ-70^\circ) = 19.7$ [15, 16, 33]				

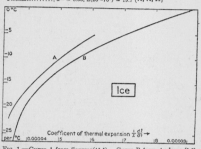

FIG. 1.—Curve A from Sawyer[43.5]. Curve B from Andrews[2.5].

II. HALOGEN COMPOUNDS

	T, °K	α	
CaF_2: Fluorite [54, 55]			$\alpha(50^\circ-60^\circ) = 5.734$ [40]
	265.7	18.5a	$\alpha^{45} = 19.34$ [56]
	235.6	17.5s	$\alpha^{40} = 19.11$ and $\Delta(20^\circ-70^\circ) = 28.8$ [21, 22, 33]
	231.4	16.7s	
	209.8	16.0a	
	186.9	14.6s	
	156.3	13.0a	
	124.9	10.2s	
	94.4	7.17	
NH_4Cl	$d_4^{20} = 1.5256$ [26].		$\alpha^{40} = 62.55$ and $\Delta(20^\circ-70^\circ) = 297.5$
$PbCl_2$	(Prefused) $d_4^0 = 5.899, a = 8.8 \ (0^\circ-50^\circ) \ [5]; cf.$ [47]		
$AgCl$	$\alpha^{40} = 32.94, \Delta(20^\circ-70^\circ) = 122.8$ [19, 20, 33]		
$PrCl_3$	$d_4^{25} = 4.0203$ [5]		
$MgCl_2.6H_2O$	$d_4^t = 1.5907$ [28]		
$CaCl_2.6H_2O$	$d_4^t = 1.7182$ [28]		
$CaCl_2.2MgCl_2.6H_2O$	$d_4^t = 1.6655$ [28]		
$LiCl$	$d_4^{25} = 2.0678$ [25]		
$NaCl$	$d_4^t = 2.1680, a = 11.2, b = 0.5 \ (-184^\circ \text{ to } 50^\circ)$ [6, 7, 12, 27, 45]; $cf.$ [19, 20, 23, 40]; $\alpha(20^\circ \text{ to } 80^\circ K) = 10.8$ [34]		
KCl	$d_4^0 = 1.9920, a = 10.5, b = 0.4 \ (-184^\circ \text{ to } 70^\circ)$ [27]; $cf.$ [19, 20, 23, 40]. $d_4^{20} = 1.9786$ (large crys.) $= 1.9841$ (small crys.) [29, 30]		
$RbCl$	$d_4^t = 2.805r, a = 12.6, b = -6 \ (0^\circ-50^\circ)$ [6]		
$CsCl$	$d_4^t = 3.9887, a = 15.9, b = -4 \ (0^\circ-70^\circ)$ [6]		
$KClO_3$	$d_4^t = 2.3467, a = 18, b = -3 \ (0^\circ-44^\circ)$ [7]		
$LiClO_4$	$d_4^{25} = 2.4284$ [42]		
$PbBr_2$	$d_4^t = 6.676, a = 9.5, (0^\circ-50^\circ)$ [5]		
$AgBr$	$\alpha^{40} = 34.7, \Delta(20^\circ-70^\circ) = 38.3$ [19, 20, 33]		
$CaBr_2$	$d_4^{25} = 3.3535$ [41]		
$NaBr$	$d_4^t = 3.213, a = 12, b = 0.3 \ (-184^\circ \text{ to } 50^\circ)$ [6, 27]		

† α' along bisector of acute angle, α'' along bisector of obtuse angle formed by optic axes, α''' along the normal to the plane of optic axes.

FIG. 3. *International Critical Tables.*

ABBÉ REFRACTOMETER. A device for the direct determination of the refractive indices of liquids.

It employs the principle of total internal reflection in a small quantity of the liquid under test which is placed between the diagonal faces of two polished right-angled prisms of dense flint glass. Observations are made by a telescope moving over a graduated scale which gives a direct reading of refractive indices. The telescope is set so that its cross wires are upon a clear line of demarcation of difference of intensity in the field of view.

The prisms are enclosed in a water jacket to allow control of temperature.

The scale may also be calibrated to read concentrations of sugar solutions since the refractive index of a solution is a function of its concentration.

J.A. FARNHAM and G.W. CANNING

ABEL EQUATION. Abel's equation is the integral equation

$$\int_0^x \frac{\varphi(s)\,ds}{(x-s)^p} = f(x)$$

where $f(x)$ is known, $f(0) = 0$, $0 < p < 1$ and it is desired to find the function $\varphi(s)$. The solution is

$$\varphi(x) = \frac{\sin p\pi}{\pi} \int_0^x \frac{f'(s)\,ds}{(x-s)^{1-p}}.$$

I.N. SNEDDON

ABERRATION OF LENS SYSTEMS. According to the principles of geometrical optics an ideal optical system should form point images of point objects and deviations from this behaviour are termed aberrations. Since the wavefronts (or surfaces of constant optical path) are orthogonal to the rays an aberration may be considered as either non-concurrence of a ray-bundle or non-sphericity of a wavefront.

Take a rectangular co-ordinate system (Fig. 1) with origin A at the centre of the exit pupil, z-axis along the principal ray and y-axis on the meridian or tangential plane (the plane containing the optical axis and the principal ray). Let S be a spherical surface, the reference sphere, with centre O on the z-axis and radius OA. The point O is the image point to which the aberrations are referred; it may be the Gaussian image point or the intersection of Oz, with any chosen image surface. Let Σ be the wavefront of the imaging

pencil which passes through A. Then the *wavefront aberration* W at a point P on the wavefront can be defined as the optical path length between Σ and S, measured along a ray and taken as positive when Σ

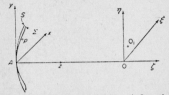

Fig. 1. To illustrate the definitions of aberrations.

is in advance of S as in the diagram. So defined, W is a function of the position of P and of the obliquity of the principal ray Oz.

Take rectangular axes $O\xi\eta$ in the image plane as shown and let the ray through P meet the $\xi\eta$-plane in O_1, with co-ordinates (ξ, η). Then (ξ, η) are the components of the *transverse ray aberration* for the ray in question.

The components of *angular ray aberration* may be defined in terms of the projections on the (x, z) and (y, z) planes of the angle between the ray PO_1, and the line PO, a normal to the reference sphere. When the aberrated ray intersects the principal ray we can define a longitudinal ray aberration as the distance between this intersection point and the image plane.

It can be seen that in every case the magnitudes of the aberrations depend on the position of the image plane. When the principal ray does not coincide with the optical axis the image plane is often taken as the Gaussian image plane and is thus oblique to the principal ray. Also the centre of the reference sphere, i.e. the origin for measurement of transverse ray aberrations, is often taken as the Gaussian image point.

If W is expressed as a function of x and y, the co-ordinates of P, the transverse aberration is given in terms of W to a certain order of approximation by

$$\left.\begin{aligned} \xi &= \frac{R}{n}\cdot\frac{\partial W}{\partial x} \\[2mm] \eta &= \frac{R}{n}\cdot\frac{\partial W}{\partial y} \end{aligned}\right\} \tag{1}$$

See Index for location of terms not found in this volume

FIG. 4. *Encyclopaedic Dictionary of Physics.*

Prof. 1961–; Chief Phys., Parkland Memorial Hospital, Dallas, Texas, 1957–; Phys., Radiation and Medical Research Foundation of The Southwest, Fort Worth, Texas, 1958–; Phys., The Tom Bond Radiol. Group, Fort Worth, Texas, 1958–; Radiol. Phys., Veterans Administration Hospital, Dallas, Texas, 1957–; Phys., Harris Hospital, Fort Worth, Texas, 1959–; Phys., St. Paul's Hospital, Dallas, Texas, 1960–; Consultant Phys., U.S. Naval Hospital, Portsmouth, Virginia, 1955–; Consultant Phys., Brooke Army Medical Center, Fort Sam Houston, Texas, 1960–; Consultant Phys., Lackland Air Force Base, Texas and Keesler Air Force Base, Mississippi; Director, American Assoc. of Physicists in Medicine; Certified in Radiol. Phys., Amer. Bd. of Radiol., 1950; Certified in Health Physics, American Bd. of Health Physics, 1960. Member, National Com. on Radiation Protection, 1961–. *Books:* Contributor to Clinical Use of Radioactive Isotopes (Fields and Seeds, 1957 and 1960); contributor to Handbook of Radiation Hygiene (McGraw-Hill, 1959). *Societies:* American Roentgen Ray Soc.; American Coll. Radiol.; Radiol. Soc. of North America; American Phys. Soc.; Radiation Res. Soc., Biophysical Soc.; Soc. Nucl. Med.; Southern Med. Soc.; Texas Radiol. Soc.; Dallas-Fort Worth Radiol. Soc.; Health Physics Soc.
Nuclear interests: Interactions of radiation with matter; Biological effects of radiation; Radiation dosimetry; Clinical uses of radioisotopes; Radiation protection; Health physics.
Address: Department of Radiology, University of Texas, Southwestern Medical School, 5323 Harry Hines Blvd., Dallas 35, Texas, U.S.A.

KROHNE, Theodore F. *Born* 1915. *Educ.:* Elmhurst College, Northwestern Univ. At present Staff Asst., Public Information, Argonne National Lab. *Societies:* American Nucl. Soc.; Nucl. Energy Writers' Assoc.
Nuclear interests: Assisting the lay public in understanding science in general and nuclear science in particular.
Address: Public Information Office, Argonne National Laboratory, Argonne, Illinois, U.S.A.

KROLL, Bruno, Ing. Personal Member, Study Assoc. for the use of Nucl. Energy in Navigation and Industry.
Address: c/o Eisenkonstruktion B. Kroll, 48 Eiffestrasse, Hamburg 26, Germany.

KROLL, Norman, Prof. Theoretical Physics Research (mainly Fundamental Particle Research) (U.S.A.E.C. contract), Columbia Univ.
Address: Nevis Cyclotron Laboratories, Irvington, New York, U.S.A.

KROLZIG, A., Dipl.-Ing. Steuerung und Regelung, Deutsches Elektronen-Synchrotron.
Address: Deutsches Elektronen-Synchrotron, 56 Flottbeker Drift, Hamburg-Gr. Flottbek 1, Germany.

KROMER, Carl-Theodor, Dr.Ing., Prof. Vorsitzer, Vorstand der Badenwerk A.G. Member, Comite d'Etudes de l'Energie Nucléaire, Union Internationale des Producteurs et Distributeurs d'Energie Électrique; Member Beirat für Kernenergie, Wirtschaftsministerium Baden-Württemberg.
Address: 19 Günterstalstrasse, Freiburg/Br, Germany.

KROMHOUT, Robert Andrew, B.S. (Phys.) (Kansas State Coll.), M.S. (Phys.) (Illinois), Ph.D. (Phys.) (Illinois). *Born* 1923. *Educ.:* Kansas State Coll.; Illinois Univ. Asst. Prof., Illinois Univ., 1954–56; Asst. Prof. (1956–60); Assoc. Prof. (1960–), Florida State Univ. *Society:* American Phys. Soc.
Nuclear interest: Nuclear magnetic resonance.
Address: Physics Department, Florida State University, Tallahassee, Florida, U.S.A.

KRONACKER, Baron P. Administrateur, Association Belge pour le Développement Pacifique de l'Energie Atomique; Membre de la Chambre des Représentants.
Address: 31 Rue Belliard, Brussels, Belgium.

KRONAUER, E. Directeur-gén. des Ateliers de Secheron S.A., Geneva. Member, Conseil d'Administration, Energie Nucléaire, S.A.
Address: 14 Avenue de Secheron, Geneva, Switzerland.

KRONBERGER, Hans, B.Sc., Ph.D. *Born* 1920. *Educ.:* King's College, Durham Univ.; Birmingham Univ. Head of Lab., Capenhurst, U.K.A.E.A., 1953; Chief Physicist, U.K. A.E.A., Risley, Director of Research and Development, U.K.A.E.A., Industrial Group, Risley, 1957; Deputy Managing Director, Reactor Group, U.K.A.E.A., Risley, 1961–. *Society:* Fellow, Inst. of Physics.
Nuclear interest: Development of prototype reactors.
Address: U.K.A.E.A., Reactor Group, Risley, nr. Warrington, Lancs., England.

KRONE, Ralph W., B.S., M.S., Ph.D. *Born* 1919. *Educ.:* Antioch College; Illinois, Johns Hopkins Univs. Assoc. Prof. Physics, Univ. Kansas. *Society:* American Phys. Soc.
Nuclear interest: Low energy nuclear physics.
Address: Department of Physics, University of Kansas, Lawrence, Kansas, U.S.A.

KRONIG, Ralph, B.A., M.A., Ph.D. *Born* 1904. *Educ.:* Columbia Univ., New York. Lecturer, Columbia Univ., 1925–27; Asst. Prof., Imperial College of Science and Technol., 1928–29; Lecturer, Groningen Univ., 1930–39; Prof. Theoretical Physics, Technol. Univ. Delft, 1939. *Books:* Band Spectra and Molecular Structure (1930); The Optical Basis of the Theory of Valency (1935); Leerboek der Natuurkunde (1946); Textbook of Physics (1954). *Societies:* Royal Netherlands Acad. of Sciences; Royal Norwegian Soc. of Sciences; Royal Inst. of Eng. of the Netherlands; Netherlands Phys. Soc.; American Phys. Soc.
Nuclear interest: Nuclear theory.
Address: Oostsingel 204, Delft, The Netherlands.

KROPIN, A. A. *Papers:* Co-author, Startup of a Cyclotron with a Spatially Varying Magnetic Field (letter to the Editor, Atomnaya Energiya, vol. 6, No. 6, 1959); co-author, A Cyclotron with a Spatially Varying

FIG. 5. *Who's Who in Atoms.*

1,151,601 <u>Optical light amplifier</u> for selective fluorescence, containing a fluorescent crystal suitable for inverse metastable energy level operation, in which the shape of the crystal is such that spontaneous or stimulated fluorescent light emitted by it is reflected back only by total reflected back only by total reflection. 6.9.61 (25.10.60 USA) WESTERN ELECTRIC CO. INC.

1,151,671 <u>Recording instrument</u> having as a stylus a capillary tube supplied with ink from a storage container through a thin hose. A rapidly releasable plug-in coupling with tapered coupling surfaces is arranged between the capillary tube and the hose. 29.1.58 DREYER, ROSENKRANZ & DROOP. A.G.

1,151,672 <u>Gyroscopic integrator.</u> A gyroscope rotor is guided in a frame which is supported in a yoke so as to be rotatable about a smaller axis at right angles to the rotor axis, the frame being rotatable about a greater axis disposed at right angles to the smaller axis. A variable signal moment about the smaller axis is fed to the gyroscope so as to generate a precession movement about the greater axis and thereby actuating an output device. A blocking mechanism prevents the precession movement until the signal moment exceeds a predetermined value. 18.10.60 (19.10.59 USA) GEN. ELECTRIC CO.

1,151,673 <u>Compound flow meter</u> for liquids. A pressure-relieved change-over valve has a lip seal which, on its side directed towards the valve seat, is flexible and which, in the closed position, engages the valve seat without friction and without guidance. 27.7.57 BOPP & REUTHER. G.m.b.H.

1,151,674 <u>Compound flow meter</u> for liquids, in which the auxiliary meter runs continuously as long as the main meter is running. An auxiliary valve has as a closure member a ball (7) arranged between two valve seats (5,6), one valve seat (6) serving as a non-return valve. 10.2.61 BOPP & REUTHER. G.m.b.H.

1,151,675 <u>Gas separator</u> A float serves as a control member for the closure device for the air outlet opening, and it is rotatable and serves also as a measuring member for the liquid flow. 17.3.58 ESTERER. FA. DR.-ING. U.

1,151,676 <u>Withdrawal device</u> for withdrawing from a storage container measured quantities of solid, granular or powdered substances, esp. such which are difficultly pourable. The material transferred into the dosing chamber is fed at an angle to its inflow direction to an outlet disposed externally of its banking angle. The material is tranferred by means of a pressurised gas or liquid, and the material quantity is controlled by variation of the pressure and/or the fluid quantity per unit of time. 7.3.61 BARTSCH. DIPL.-ING. DR.-ING. W.

1,151,678 <u>High-power variable objective</u> consisting of a base objective on the side towards the film, and of a variable section having between two fixed members two axially slidable members. The base objective as well as the variable section consist each of four members in air. The rear member of the variable section near the base objective and the adjacent slidable member are uncemented individual lenses. The two slidable negative members have their concave surfaces directed towards one another. 16.9.60 ISCO OPT. WERKE. G.m.b.H.

1,151,679 <u>X-ray diffraction device</u> containing a crystal and a radiation detector. The X-ray fluorescent radiation generated by the tested object is spectrally resolved by the crystal and applied to the detector. The latter is fixed to a supporting arm rotating about the same axis as the crystal and at twice the angular velocity. A nut (16) is fixed on a rod (15) on the supporting arm (14) and moves on a threaded spindle (17) which rotates about a pin (12). The pin (12) and the rod (15) are at equal distance from the axis of rotation (13) of the supporting arm and crystal. The threaded spindle is connected by a transmission between the pin (12) and axis (13) of the crystal (5) and is moved at constant speed. 6.12.60 (10.12.59 USA) N.V.-PHILIPS GLOEILAMPENFAB.

1,151,680 <u>Ionisation vacuum gauge</u> in which a wire-shape collector electrode is stretched inside a grid-like anode, and a hot cathode is provided outside the latter. The potential difference between the anode and the cathode is about one fifth of that between the anode and the collector electrode The diameter of the latter is less than 10 μ. 12.7.61 (15.7.60 Neth.) N.V. PHILIPS' GLOEILAMPENFAB.

1,151,681 <u>Balancing machine</u>, in which the phase of a voltage generated by the unbalance is compared with a reference phase obtained from the balanced object which is provided with at least two marks separated by 180°. They are photoelectrically scanned, and a bistable flip-flop circuit is controlled by impulses so obtained. 11.2.59 LOSENHAUSENWERK DUSSELDORFER MASCHINENFAB. A.G

1,151,683 <u>Continuous humidity measurement</u> of pre-dried loose material such as small wooden pieces, shavings, coarse fibres, etc., which for resistance measurement are passed under pressure between two roller electrodes with gripping surfaces. One of the two electrodes (3) has a stationary support, whilst the other (5) is adjustable and spring-loaded in a plane substantially vertical to that defined by the axes of rotation of both rollers (3,5).

FIG. 6. *German Patent Abstracts.*

PHYSICS
Fundamentals

827. AHARONOV, Y. (Br). Some problems in the quantum theory of measurements and electromagnetic potentials as observables in the quantum theory. PH.D.

828. CARMI, G. (Br). On the separation between collective and individual aspects in a many-body problem, without the use of redundant variables. PH.D.

829. KERR, R. P. (C, Trinity). Equations of motion in general relativity. PH.D.

830. GERBER, G. J. (LKC). A critical investigation and extension of Eddington's methods of linking curvature and wave functions. PH.D.

831. GILSON, J. G. (LUC and BkC). Investigations in non-linear quantum electrodynamics. PH.D.

832. HUCK, R. J. (LUC). Variational methods in inelastic scattering theory. PH.D.

833. MORDUCH, G. E. (LIC). Lorentz-invariant theories of gravitation. PH.D.

834. STREATER, R. F. (LIC). Quantum theory of fields. PH.D.

835. UNDERHILL, J. (LUC). Investigations in non-relativistic quantum scattering theory. PH.D.

836. VAN DER BURG, M. G. J. (LKC). Axisymmetric solutions in general relativity. PH.D.

837. STEINER, E. (M). An introduction to the quantum mechanics of matter confined to a two-dimensional space. M.SC.

838. WESTWOOD, J. (O, Somerville). The quantum mechanics of two electron and many electron systems. B.SC.

839. YOUNG, W. H. (S). Many-particle systems in quantum mechanics. PH.D.

See also 774, 864, 1452a.

Mechanics
Statistical Mechanics

840. BOOT, A. R. (LQMC). Statistical mechanics of transport processes in fluids. PH.D.

841. BURLEY, D. M. (LKC). Some problems in the theory of co-operative phenomena. PH.D.

See also 949.

Mechanics of Fluids
General

842. MEHRABIAN, Y. (B). Flow visualisation applied to valves. M.SC.

843. NICHOLL, C. I. H. (C, Emmanuel). A study of heated turbulent boundary layers at low Reynolds numbers in air. PH.D.

FIG. 7. *Index to Theses accepted for Higher Degrees in the Universities of Great Britain and Ireland.*

CONTENTS

THE PHYSICAL REVIEW   SECOND SERIES, VOLUME 129, No. 1   1 JANUARY 1963

(Continued on cover three)

LANCASTER PRESS, INC., LANCASTER, PA.

FIG. 8. *The Physical Review.*

PHYSICS

General and Miscellaneous

Refer also to abstracts 20149 and 20166.

20534 (AD-293438) LONG LIFE CLOSED LOOP MHD RESEARCH AND DEVELOPMENT UNIT. Interim Scientific Report No. 3, September 15, 1962 to December 15, 1962. (Westinghouse Electric Corp. Research Labs., Pittsburgh). Dec. 15, 1962. Contract AF33(657)-8311. 57p.

The construction of the component parts of the magneto-hydrodynamic closed loop and its auxiliaries is discussed. Topics covered include: status of closed loop duct, status of closed loop auxiliaries, enhanced ionization utilizing chemical disequilibrium, compatibility of tantalum and cesium, and fail safe features of cesium—helium feed and bleed system. The various possible modes of failure of the overall test apparatus were analyzed and the appropriate safety provisions were included in the design. (M.C.G.)

20535 (ARL-63-29) NON-UNIFORM SCALING OF AN APPROXIMATE H_2^+ WAVE FUNCTION. William J. Meath and Joseph O. Hirschfelder (Wisconsin. Univ., Madison. Theoretical Chemistry Lab.). Nov. 1962. Contract AF33 (657)-7311. 11p.

The approximate wave function considered for the H_2^+ molecule is $\Psi = N \exp \{-s[\gamma^2(x^2 + y^2) + (z + R/2)^2]^{\frac{1}{2}}\} + \exp \{-s[\gamma^2(x^2 + y^2) + (z - R/2)^2]^{\frac{1}{2}}\}$. Energy optimization of the nonuniform scaling parameter, γ, results in the satisfaction of the "one-coordinate" virial theorem $-(\hbar^2/2m) \int \Psi(\partial^2/\partial x^2) \Psi d\tau = (\frac{1}{2}) \int \Psi x(\partial V/\partial x) \Psi d\tau$. Similarly the energy optimization of the uniform scaling parameter, s, results in the satisfaction of the usual virial theorem. The simultaneous application of these two types of scaling is also discussed. (auth)

20536 (ASD-TDR-62-320) STUDY OF MHD AND EHD FREE-CONVECTION ENERGY CONVERTERS. Meredith C. Gourdine (Plasmadyne Corp., Santa Ana, Calif.). Apr. 1962. Contract AF33(616)-8007. 110p. (AD-282325)

A theoretical and experimental feasibility study of single and two-phase free-convection MHD and EHD energy conversion is summarized. Single-phase operation is shown to be impractical because the maximum possible efficiency of converting heat into flow kinetic energy is very low. On the other hand, two-phase operation is far more efficient and definitely has possibilities for practical energy conversion applications. (auth)

20537 (ASD-TDR-63-175) STUDY OF ELECTRICAL AND PHYSICAL CHARACTERISTICS OF SECONDARY EMITTING SURFACES. W. T. Peria, ed. (Minnesota. Univ., Minneapolis). Jan. 23, 1963. Contract AF33(657)-8040. 126p.

Results are presented which show that it is necessary to sputter-clean titanium crystals in the study of the variation of secondary emission with the angle of incidence of the primary beam. The nature and cause of the errors introduced when titanium is not sputter-cleaned are described. The asymmetrical structure observed in the angular dependence studies is attributed to misalignment of the crystals and some experimental results are given as evidence. The energy dependence of the magnitude of the structure in the angular dependence curves for tungsten is examined and found to be basically the same as for titanium and germanium. It is also demonstrated that sputter-cleaning is not necessary for angular dependence studies in the case of tungsten. The techniques for production of unbacked gold films are outlined. Preliminary results of studies of the energy distribution of electrons transmitted through one of these films are reported. Further results of studies of MgO films formed by the oxidation of a Mg film on a metal substrate are presented and analyzed. The concentrated study of two particular MgO films formed on Mo and W substrates has led to a quite detailed picture of the trapping centers that can occur in MgO films. Preliminary results for a second set of films are also reported. A method involving use of the destructive interference of reflected light from anodized tantalum has been devised for the determination of the ion current density as a function of position, on the target face, during the sputter-cleaning operation. An optical arrangement for measuring the amount of sputtering as a function of position on a Ta_2O_5–Ta target was built, and preliminary measurements indicate the feasibility of the measuring technique. The difficulties encountered in the study of the dissociation of NaCl thin films by low energy electron bombardment are described. The migration of Na^+ ions to the film surface is believed to be the cause of the difficulties. Results of dissociating MgO thin films are presented, and the problem of adsorbed gas on the target surface is discussed. Slow electron diffraction from the (001) surface of a MgO crystal reveals two different types of surface structure. One is a fc structure with the MgO lattice constant. The other has a double spacing in those azimuths for which the sum of the indices is even. The effects of heating and sputtering on the pattern from the fc structure are described. The effect of these treatments on the secondary electron emission is also described. A qualitative explanation of the "extra" intensity maxima in slow electron diffraction patterns is presented. A new model is proposed to explain the major peaks observed in the curves of the variation of secondary emission from single crystals as a function of the angle of incidence of the primary beam. Qualitative explanations of the manner in which the magnitude and width of these peaks vary with primary energy are given in terms of the model. The reason for the differences in the characteristics of the peaks for metals and insulators is also suggested by the model. The small peaks that occur between and superimposed upon major peaks can also be explained if the new model is combined with parts of a model previously proposed. The disagreements between this model and some of the standard assumptions in secondary emission theories are briefly discussed. Using a previous calculation as a starting point, the influence of an internal electric field on the secondary emission from ionic crystals is computed. The results are in fair agreement with experiment for the case of MgO. 41 references. (auth)

20538 (BSD-TDR-63-60) MEASUREMENT OF LOW RESISTANCE AND THE AC RESISTANCE OF SUPERCONDUCTORS. Jacob L. Zar (Avco Corp. Avco-Everett Research Lab., Everett, Mass.). Jan. 1963. Contract AF 04(694)-33. (AD-299782; AMP-100)

The resistance of highly conductive materials is measured by an induction method that does not require a direct electrical connection to the test specimen. Instead, the test sample is used as a short circuited secondary winding of a transformer. The method is suited to specimens where lead correction errors would introduce difficulty, or where the temperature or field environment requires

FIG. 9. *Physics Abstracts.*

EXPERIMENTS ON THE OMEGATRON TYPE MASS SPECTROMETER AS A VACUUM ANALYSER. See Abstr. 22283 of 1962.

184   NON-DESTRUCTIVE LEAK CONTROL OF SMALL SEALED ENCLOSURES. H.Piatti.
Vide (France). Vol. 17, 250-5 (May-June, 1962). In French.
It is well known that the flow of a leak can be evaluated if the increase of pressure caused by this leak in a given volume during a given time is measured. Using simple means such as a U-shaped manometer with a rough vacuum it is possible to measure extremely small leaks from small sealed enclosures in a relatively short time (order of 10^{-4} mm Hg litre sec^{-1} in a few minutes). This method has the great advantage of being non-destructive and does not require a special gas inside the envelope (i.e. helium). It can

be adapted for the control in mass production of crystal cans and of any other small-volume envelopes. This method is much more economical than any other current practice method.

185   VACUUM AND ULTRAVACUUM IN AN ION - INJECTION MAGNETIC MIRROR DEVICE.
E.Fischhoff, F.Prevot and Z.Sledziewski.
Vide (France). Vol. 17, 195-200 (May-June, 1962). In French.
The vacuum system of an ion-injection magnetic mirror device is described. The tube using gold gaskets is bakable up to 400 C. Oil diffusion units give a base pressure below 10^{-9} mm Hg. These units are arranged to provide differential pumping for local gas sources. In addition ion pumping and getter pumping is used. The problem of measuring transient low pressures is briefly discussed.

VIBRATIONS . WAVES . ACOUSTICS

VIBRATIONS . ELASTIC WAVES

186   EXACT SOLUTION FOR THE VIBRATIONS OF A NONLINEAR CONTINUOUS MODEL STRING.
N.J.Zabusky.
J.math. Phys. (USA), Vol. 3, No. 5, 1028-39 (Sept. -Oct., 1962).
An exact solution is given for the partial differential equation $y_{tt} = [1 - \epsilon y_x] y_{xx}$ which describes the standing vibrations of a finite, continuous, and nonlinear string. The nonlinearity studied, $[1 - \epsilon y_x]^{\alpha}$, was motivated by the work of Fermi, Pasta, and Ulam (1955), where they reported on numerical studies of the "equipartition of energy" in nonlinear systems. To obtain the solution, the above equation is transformed into a linear equation by inverting the roles of the dependent (u = y, and v = y_t) and independent (x and t) variables. Riemann's method of integration is applied to the problem and the solutions for t and x are written as integrals. The nature of the "inverse Riemann plane," how it is related to the initial conditions, and how one unfolds it are discussed in detail. A general procedure is described for reinverting the solution, so that y can be written as a function of x and t. It is illustrated to order ϵ for the above problem. It is demonstrated that y_{xx} becomes singular, that is, y_x develops a discontinuity after an elapsed time of order (1ϵ). The methods described are applicable to any nonlinear string where the coefficient of y_{xx} is a function of y_x only. The effect of higher spatial derivatives on the formation of the singularity is discussed.

187   LONGITUDINAL VIBRATIONS OF A CONICAL BAR.
V.Vodička.
Appl. sci. Res. A (Netherlands), Vol. 11, No. 1, 13-16 (1962).
The present paper considers longitudinal vibrations of a homogeneous truncated cone. This is probably one of the few cases relating to vibrating rods of variable cross-section where it is possible to give an exact solution instead of using the well-known laborious approximate procedures. Classical methods are applied although the problem in question could be treated also by means of the Laplace transformation.

188   NONLINEAR FLEXURAL VIBRATIONS OF SANDWICH PLATES. Y.Y.Yu.
J. Acoust. Soc. Amer., Vol. 34, No. 9, 1176-83 (Sept., 1962).
A set of nonlinear equations describing the vibrations of sandwich plates is derived. These equations are then applied to a plate vibrating in plane strain and to a rectangular plate, with the boundary edges assumed to be hinged in both cases. It is shown that the nonlinear frequencies increase with the amplitude of vibration. It is shown also that transverse shear deformations play an important role; however the effect produced by these deformations decreases with increasing amplitudes of vibration.

189   HIGH-FREQUENCY EXTENSIONAL VIBRATIONS OF SANDWICH PLATES. H.N.Chu.
J. Acoust. Soc. Amer., Vol. 34, No. 9, 1184-90 (Sept., 1962).
A set of approximate equations governing the extensional motion of a sandwich plate, together with the associated initial

and boundary conditions are derived. Both the sandwich core and the facings are assumed homogeneous and isotropic, and the facings are assumed thin and identical to each other. The coupling effect of the thickness deformation of the core is considered in the derivation. Frequency–wavelength curves are then obtained from these equations for an infinite plate in plane strain. The corresponding curves for the same plate according to the three-dimensional exact theory of elasticity are also obtained. These two sets of curves are then used as a basis for discussion of the accuracy of the approximate equations. The curves indicate that for commonly used core materials and for wavelengths about three times the plate thickness and longer, the approximate equations are reasonably good up to frequencies somewhat higher than the fundamental thickness-stretch frequency of the plate.

190   VIBRATIONS OF POINT-DRIVEN CYLINDRICAL SHELLS. M.Heckl.
J.Acoust. Soc. Amer., Vol. 34, No. 10, 1553-7 (Oct., 1962).
By using some approximations it is possible to get fairly simple approximate formulae for the number of resonance frequencies and for the point impedance of thin cylinders. Measurements of the resonance frequencies and of the number of resonances within a certain frequency range confirm the theoretical results.

191   TAPERING A BAR FOR UNIFORM STRESS IN LONGITUDINAL OSCILLATION. D.A.Bies.
J.Acoust. Soc. Amer., Vol. 34, No. 10, 1567-9 (Oct., 1962).
A cross-sectional taper is described which provides a bar with the possibility of uniform stress in longitudinal oscillation. It is shown that this kind of motion is a special case of a more general kind of motion in which it is possible by properly terminating the bar to make the displacement and the stress functions related polynomials of any desired order. A theory which describes the behaviour of the bar in longitudinal motion is presented and compared with the results of experiment.

192   TORSIONAL VIBRATION OF BEAMS HAVING RECTANGULAR CROSS SECTIONS. M.Vet.
J. Acoust. Soc. Amer., Vol. 34, No. 10, 1570-5 (Oct., 1962).
An elastic energy method is presented for finding the fundamental torsional vibration frequency of solid rectangular beams without boundary constraints. A Rayleigh–Ritz method is employed, and the results, which were calculated on the IBM 650 computer, are shown. Also illustrated is the effect a change in the Poisson's ratio of the beam material has on the fundamental frequency. A comparison is made with results of torsional vibration tests performed at the National Bureau of Standards. This comparison indicates that the analytical method employed gives fairly accurate results.

193   PERIOD EQUATION FOR RAYLEIGH WAVES IN A LAYER OVERLYING A HALF SPACE WITH A SINUSOIDAL INTERFACE. J.T.Kuo and J.E.Nafe.
Bull. Seismol. Soc. Amer., Vol. 52, No. 4, 807-22 (Oct., 1962).
The problem of the Rayleigh wave propagation in a solid layer overlying a solid half space separated by a sinusoidal interface is investigated. The amplitude of the interface is assumed to be small.

FIG. 10. *Nuclear Science Abstracts.*

Section II

MATHEMATICAL AND PHYSICAL

BRITISH ASTRONOMICAL ASSOCIATION. Registered Office: 303, *Bath Road, Hounslow West, Middlesex.* London Headquarters: *Burlington House, W.1.* (Founded 1890.)
Objects: (1) The association of observers, especially the possessors of small telescopes for mutual help, and their organization in the work of astronomical observation; (2) the circulation of current astronomical information; (3) the encouragement of a popular interest in astronomy.
Membership: Numbers 2,600. Annual subscription: £2 5s. 0d.; Entrance fee, 5s.; Student membership, under 25, £1 10s. 0d.
Meetings: Burlington House, W.1. Last Wednesday in month, October to June.
Publications: *Journal*, 8 times a year. *Annual Handbook. Memoirs* (irregular).
Library: Burlington House, W.1. 3,500 volumes.

BRITISH COMPUTER SOCIETY, *Finsbury Court, Finsbury Pavement, London, E.C.*2 (Telephone: MONarch 6252.) (Founded 1957.)
Object: To further the development and use of computational machinery and techniques related thereto.
Membership: Annual subscription: Ordinary Members, £3 3s. 0d., Entry fee, £1 1s. 0d. Associate Members: over 21 and under 25, £1 1s. 0d.; No entry free. Institutional Members: Annual subscription, £10 10s. 0d.; No entry fee.
Meetings: Lectures, Study Groups and Colloquia regularly from September to April in London and provinces.
Publications: *Journal* and *Bulletin.*

BRITISH HOROLOGICAL INSTITUTE, 35, *Northampton Square, London, E.C.*1 (Telephone: CLErkenwell 4413.) (Founded 1858.)
Objects: To promote the cultivation of the science and to teach the practice of Horology.
Membership: Numbers, 4,000. Annual subscription: £2 2s. 0d.
Meetings: Last Thursday in each month, from October to April, inclusive. Annual General Meeting: October.
Publication: *The Horological Journal* (monthly), 18s. per annum, free to Members.

THE BRITISH INSTITUTE OF RADIOLOGY, 32, *Welbeck Street, London, W.1.* (Telephone: WELbeck 6237 and 6867.) (Founded 1927.)
Objects: The Institute is designed to serve as a meeting place for radiologists, physicists, X-ray engineers and manufacturers, and radiographers; to form a centre for consultation and co-ordination of the medical, physical and biological aspects of radiology and X-ray engineering problems; to provide a bureau of information for London, provincial overseas and foreign workers; to promote the advancement and study of radiology radio-activity and allied subjects.
Membership: Numbers approximately 1,350. Membership subscription: Home, £5 5s. 0d. per annum; Overseas, £4 4s. 0d. per annum.
Meetings: Monthly.
Publications: *British Journal of Radiology* and Supplements. Subscription, £5 15s. 6d. per annum; Supplements extra.
Library: Library containing current textbooks and journals on radiology and allied subjects to which continual additions are made.

BRITISH INTERPLANETARY SOCIETY, 12, *Bessborough Gardens, London, S.W.*1. (Telephone: TATe Gallery 9371.) (Founded 1933.)
Objects: To promote the development of astronautics by the study of rocket engineering, astronomy and associated sciences.
Membership: Numbers 3,500. Annual subscriptions: Fellows, £4 4s. 0d.; Associate Fellows, £3 13s. 6d.; Members, £2 12s. 6d.; Corporate Members, £100 p.a.
Meetings and Symposia: Several in each month, from October to April inclusive, with additional special symposia during the summer months. Annual General Meeting: September. Regular Meetings held also at Branches in Birmingham, Bristol, Derby, Glasgow, Leeds, Manchester, South Shields, Washington, D.C., U.S.A., and three in Australia.
Publications: *Journal* (bi-monthly) and *Spaceflight* (bi-monthly), £2 10s. 0d. to Libraries only (free to Members and Fellows).
Library: Contains a specialized collection of works on rocket engineering and space flight.

BRITISH OPTICAL ASSOCIATION, 65, *Brook Street, London, W.1.* (Telephone: MAYfair 3382.) (Founded 1895.)
Objects: (1) To encourage the science of Optics and its application to the improvement.

FIG. 11. *Scientific and Learned Societies of Great Britain.*

scientific and engineering knowledge relating to plastics.

Membership: Senior, those qualified by previous technical training, experience, or present occupation to conduct or direct design, engineering, chemical or physical research relating to plastics, to exercise technical supervision of the production of plastics materials or products, or the manufacture of equipment connected therewith, to impart technical instruction in the chemistry, physics or engineering of plastics, or the design and fabrication of plastics products; Associate, those qualified as Senior Members except having less experience qualification, or engaged in the plastics or a related industry in a responsible commercial, financial, or manufacturing capacity, or connected with the plastics or a related industry and competent to cooperate technically with plastics engineers and scientists; Junior Member, those qualified to fill subordinate technical positions in the plastics or a related industry, and less than thirty years of age; Student Member, those regularly enrolled in a technical course at a recognized college or university, and less than twenty-six years of age; Distinguished Member, those who, at the time of election, are members in good standing who have attained professional eminence in the plastics field, who have made a significant contribution to the plastics industry, who have given outstanding devotion and service to the Society, or who have served the Society as President; Honorary Member, those who have been non-members but who, by virtue of outstanding achievement or professional eminence, may be deemed worthy of this status. Total membership 7,679.

Meetings: Annual; 7 Regional; 400 Sectional.

Publications: Journal, monthly, current volume: 16, $6, domestic, $10 foreign. Editor: Melvyn A. Kohudic.

1375. Society of Protozoologists. *President:* William Trager, The Rockefeller Institute, New York 21, N. Y. Term expires December 1961. *Secretary:* John O. Corliss, Department of Zoology, University of Illinois, Urbana, Ill. Term expires September 1962.

History: Organized December 1947.

Purpose: To promote closer association of persons interested in research which will advance protozoological science.

Membership: Open to any individual, including non-Americans, interested in the science of protozoology. Regular Members; Graduate Student Members; Honorary Members, 10, persons who have rendered highly meritorious service in the field. Total membership 675.

Meetings: Annual.

Publications: Journal of Protozoology, quarterly, current volume: 7, $12, free to members. Editor: William Trager.

1376. Society of Public Health Educators, Inc. *President:* Norbert Reinstein, 153 East Elizabeth Street, Detroit 1, Mich. Term expires October 1960. *Executive Secretary:* Ruth F. Richards, 42 Broadway, Room 900, New York 4, N. Y. Term expires October 1961.

History: Incorporated 1952.

Purpose: To contribute to the advancement of the health of all people by improving standards in the field of public health education.

Membership: Two years experience and master's degree in public health. Total membership 360.

Meetings: Annual.

Publications: Monograph, quarterly, current volume: 8, $3. Editor: H. Weddle.

1377. Society of Rheology. c/o American Institute of Physics, 335 East 45th Street, New York 17, N. Y. *President:* John H. Elliott, Hercules Powder Company, Research Center, Wilmington, Del. Term expires November 1961. *Secretary:* William R. Willetts, Titanium Pigment Corporation, 99 Hudson Street, New York 13, N. Y. Term expires November 1961.

History: Organized December 1929. Member society of the American Institute of Physics.

Purpose: The development of the science of the deformation and flow of matter.

Membership: Open to chemists, physicists, engineers, and other professional scientists. Personal Members, 421; Company and Institutional Members, 30; Sustaining. Total membership approximately 550.

Meetings: Annual.

Professional activities: Bingham Medal, award annually to a scientist who has made a notable contribution to rheological knowledge.

Publications: Rheology Bulletin, three times yearly. Transactions, annual, current volume: 4, both free to members. Editor: E. H. Lee, Department of Applied Mathematics, Brown University.

1378. Society of Soft Drink Technologists. 1128 16th Street, N. W., Washington 6, D. C. *President:* Melvin Helin, Infilco Inc., Tucson, Ariz. Term expires April 1961. *Secretary-Treasurer:* Harry E. Korab. Term indefinite.

History: Organized July 1953; incorporated 1954.

FIG. 12. *Scientific and Technical Societies of the United States and Canada.*

Siebe, Gorman & Co., Ltd., Neptune Works, Davis Road, Chessington, Surrey
Laboratories : As above
Scope of Research : Diving apparatus, aqualungs, underwater equipment, breathing apparatus, gas and dust respirators, protective clothing, resuscitation apparatus, safety helmets and aircraft seat belts
Chief Development Engineer : L. R. Phillips

Simon Engineering Ltd., Cheadle Heath, Stockport, Cheshire
HENRY SIMON LTD.
Subjects of Research : Cleaning, conditioning, cutting, grinding, grading, mixing, extruding, baking of cereal and cereal products, for the flour, feed, food and bakery industries ; pneumatic and mechanical handling, instrumentation and control of processes ; cardboard box making machinery and plastic coating of materials ; gear design and transmission
Head of Research and Development : R. W. Allen, M.Sc., A.M.I.Mech.E., Mem.A.S.M.E.
Head Physicist : T. J. Brown, B.Sc., A.Inst.P., A.R.C.S.
Head Chemist : B. H. Wragg, M.A.(Cantab.)
Senior Development Engineer : F. Shoesmith, B.Sc.
Number of Qualified Staff : 15   *Floor Space :* 30,000 sq. ft.

SIMON-CARVES LTD.
Subjects of Research : Coal preparation ; coal carbonisation and by-products ; effluent treatment ; gas purification ; corrosion ; flow of materials ; application of radio-isotopes ; particle mechanics ; micro-biology ; methods of analysis
Director of Research : T. Kennaway, B.Sc., A.M.I.Chem.E., A.R.I.C.
Research Manager : W. Bostock, B.Sc.
Head of Physics Section : L. Cohen, B.Sc., Ph.D., F.Inst.P.
Head of Chemical Research : K. H. Todhunter, M.Sc., A.M.I.Chem.E.
Head of Chemical Testing : G. E. Preston, B.Sc., F.R.I.C.
Metallurgist : A. C. Harris, Assoc.Met., F.I.M.
Number of Qualified Staff : 20   *Floor Space :* 30,000 sq. ft.

Slumberland Group Ltd., Redfern Road, Tyseley, Birmingham, 11
Laboratory : Slumberland (Research) Ltd., 547 Buxton Road, Great Moor, Stockport, Cheshire
Subject of Research : Bedding and furniture
Research Director : T. C. Williams, M.Sc., F.Inst.P.
Floor Space : 15,000 sq. ft.

Smith & Nephew Associated Companies Ltd., 2 Temple Place, Victoria Embankment, London, W.C.2
Laboratories : Smith & Nephew Research, Ltd., Hunsdon Laboratories, Ware, Herts.
Scope of Research : Biology, pharmacy, physics and physical, analytical, organic and technical chemistry in the fields of surgical dressings, pressure sensitive adhesives, cosmetic and toilet preparations, ethical pharmaceutical preparations and in connection with certain kinds of textile goods and cellulose and cotton fibre products for personal use
Directors : D. E. Seymour, F.R.I.C.; E. M. Bavin, B.Sc., F.R.I.C. ; D. Suddaby, B.Sc., F.R.I.C.
Number of Qualified Staff : 50

FIG. 13. *Industrial Research in Britain.*

FÖRSVARETS FORSKNINGSANSTALT, AVDELNING 2

The Research Institute of National Defence, Department of Physics

Affiliation: Department of the Institute.

Linnégatan 89, Stockholm 80, Sweden - Telephone:
Stockholm   63 15 00

Head: T. Magnusson

Research activities: Research in physical fields for defence purposes.

LU, FYSISKA INSTITUTET

LU, Institute of Physics

Affiliation: Institute of the University of Lund, see p. C. 43.

Sölvegatan 14, Lund, Sweden - Telephone: Lund   172 30

Head: Sten von Friesen - 5 scientists.

Research activities

cover atomic spectroscopy and nuclear physics and electronics. Precise mass determinations on heavy mesons and hyperons and other cosmic ray particles, energy measurements on gamma radiation by means of scintillation spectrometers, studies of continuous gamma- and x-radiations, corona discharges and precision measurements of electronic phenomena. 1200 MeV electron synchrotron is being built.

FIG. 14. *Scandinavian Research Guide.*

24. Pharmacology and Therapeutics (*continued*)

P.T. Nowell, Ph.D.   The thymus gland.   Metabolism of drugs
A. Knifton   The action of drugs on the pig uterus

25. Physics

Experimental:

Professor J.M. Cassels,   High energy physics
  F.R.S.
*Professor A.W. Merrison
B. Collinge, Ph.D.   High energy physics using 380 MeV cyclotron
W.H.R.F. Muirhead,
  Ph.D.
J.R. Wormald, Ph.D.
S.G.F. Frank, Ph.D.
A.N. James, Ph.D.
G.R. Court, Ph.D.
M.J. Moore   Plant engineering
L.L. Green, Ph.D.   Nuclear physics, using tandem Van de Graaff generator
J.C. Willmott, Ph.D.
G. Parry, Ph.D.
P. Dagley, D.Phil.
P.T. Andrews, Ph.D.
I.G. Main, Ph.D.
W.H. Evans, Ph.D.   Liquid hydrogen bubble chambers
D.N. Edwards, Ph.D.
P. Mason, Ph.D.
R.A. Donald, Ph.D.
J.R. Holt, Ph.D.   Electron synchrotron design
H.D. Parbrook, Ph.D.   Acoustics
M.G. Davies, Ph.D.
W. Tempest, Ph.D.
R. Holmes, Ph.D.
D.G.E. Martin β-ray spectroscopy

Theoretical:

Professor H. Fröhlich,   Theory of superconductivity. Field theory
  D.Sc., Dr.Phil., F.R.S. Theory of solids
R. Huby, Ph.D.   Nuclear physics
G.R. Allcock, Ph.D.   Quantum field theory
H.C. Newns, Ph.D.   Nuclear physics
G. Rickayzen, Ph.D.   Theory of superconductivity

26. Physiology

Professor R.A. Gregory,   Gastric secretion
  D.Sc., Ph.D.
T.G. Richards, M.D.,   Circulation in special regions of the body
  Ph.D.
E.R.L. O'Maille
A.H. Short
I. Calma, M.D.   Neurophysiology
G. Kidd, Ph.D.
D.V. Roberts, M.D.
H.J. Tracy   Gastric secretion
A.G. Singleton   Ruminant physiology
K.D. Neame, Ph.D.   Amino-acid transport

*Director, National Institute for Research in Nuclear Science Electron Laboratory

Fɪɢ. 15. *Scientific Research in British Universities and Colleges, 1962-63.*

HYDROMECHANICS

British Hydromechanics Research Association

South Road
Temple Fields
Harlow, Essex
Tel.: Harlow 24366/8

Director: L. E. Prosser, B.Sc.(Eng.), A.K.C., M.I.Mech.E., M.I.W.E.
Information Officer: H. S. Stephens, M.I.Inf.Sc., A.M.I.E.D.

SCOPE OF WORK

Fundamental fluid mechanics; centrifugal and allied pumps; fans and blowers; control and motion of fluids in pipes and open channels; model studies of hydraulic structures; hydraulic transmission of power; seals and glands; fluid power and oil hydraulics; automatic control and servomechanisms.

THE LIBRARY

Lending and reference facilities are available free of charge to members and to bona-fide workers who are non-members.

The library gives comprehensive coverage, by books, journals, abstracts and published and unpublished reports, of fluid mechanics and hydraulic engineering. In addition references are collected from literature on mechanical engineering, civil engineering, chemical engineering, mining, ventilation, mathematics, mechanics, information and library materials and ehotography.

FOREIGN PUBLICATIONS IN LIBRARY

The library holds a selection of books, journals and abstracts on fluid mechanics and hydraulic engineering principally from U.S.A., Germany, Austria, Switzerland, U.S.S.R., France, Italy, Japan, Scandinavia and Hungary.

PUBLISHED BY THE ASSOCIATION

The *Abstract Bulletin* (issued bi-monthly) is free to members and available to universities and technical libraries on an exchange basis. Non-members may purchase it (ordinary subscription rates U.K. £3 3s., elsewhere £3 12s. or $10.00 per volume). The Association issues to members *Research Reports, Technical Notes, Design Guides*, reviews and translations as they become available, but some older publications are available to non-members at a small charge.

INQUIRY SERVICE

Advice is given normally without charge to members, and a consultation fee is charged to non-members. Advice is given by correspondence, telephone and interviews.

OTHER INFORMATION WORK

The Association holds open days to which selected non-members are invited. There is also a mobile exhibition and demonstration unit which visits members' works. Residential conferences are held for members; there is a charge for accommodation.

CONSULTATIVE SERVICES

The Association's services to members cover hydraulic problems of all kinds: flow in machines; flow in pipes, passages and open channels; hydraulic servo-mechanisms, seals; pressure surges; cavitation; hydraulic transport of solid materials; hydraulic model studies of civil engineering structures; testing of all types of fluid machines.

FIG. 16. *Technical Services for Industry.*

Dynamics, Oscillations and Gyroscopes
[Dewey 531.3, 531.32, 531.34 Classes]

Dynamics is the study of the behaviour of objects in motion and in particular the study of objects acted on by forces and having variable velocity. The foundations of this branch of mechanics are Newton's three laws of motion. The following is a selection of some of the more recent textbooks:

ATKINS, R. H., *Classical Dynamics*, Heinemann, London, 1959.

GLAUERT, M. B., *Principles of Dynamics*, Routledge, London, 1960.

SHORT, A. E., *Dynamics*, University of London Press, London, 1959.

WHITTAKER, E. T., *A Treatise on the Analytical Dynamics of Particles and Rigid Bodies*, Cambridge University Press, London, 4th ed., 1960.

Oscillations are vibratory or periodic motion. Their theory can be followed in the book by W. G. Bickler and A. Talbot, *An Introduction to Vibrating Systems*, Oxford University Press, London, 1961; or in D. S. Jones's treatment of systems with one degree of freedom, *Electrical and Mechanical Oscillations*, Routledge, London, 1961; or in R. F. Chisnell's *Vibrating Systems*, Routledge, London, 1960. This deals with the vibration of linear mechanical and electrical systems with two degrees of freedom.

Gyroscopes are used in navigational systems and consist basically of a well-balanced disc which is free to rotate about an axis itself confined within a framework so that it is free to take any orientation. When spinning, the axis of rotation remains fixed, regardless of the motion of the outer framework. Much development work is being done by government laboratories and is shrouded in secrecy so that the present possible accuracy of the system is not generally known. The theory and design of gyroscopes can be studied using sources such as:

ARNOLD, R. W. and MAUNDER, L., *Gyrodynamics*, Academic Press, New York, 1961.

D

SAVET, P. H., *Gyroscopes: Theory and Design*, McGraw-Hill, New York, 1961.

Elasticity, Rheology [Dewey 531.38 Class]

Elasticity is that property of a body which allows it to automatically recover its normal configuration as the deforming forces are removed. The mathematical theory of elasticity is covered by an early standard work by A. E. H. Love, *A Treatise on the Mathematical Theory of Elasticity*, Cambridge University Press, London, 4th ed., 1927, and by J. C. Jaegar's fairly elementary account *Elasticity, Fracture and Flow*, Methuen, London, 2nd ed., 1962.

Several Russian authors have published in this field. L. D. Landau and E. M. Lifshitz cover the theory of deformation, conductivity and viscosity of solids and some problems concerning elastic plates and shells in their *Theory of Elasticity*, Pergamon, London, 1959. V. V. Novozhilov's *Theory of Elasticity*, Pergamon, London, 1961, develops a completely general method of approach to any problem in the theory of elasticity and then shows how, on certain conditions, the general equations can be modified to the more familiar linear equations.

Other recommended texts include:

GOODIER, J. N. and HODGE, P. G., *Elasticity and Plasticity*, Wiley, New York, 1958.

GREEN, A. E. and ZENA, W., *Theoretical Elasticity*, Clarendon Press, Oxford, 1954.

TIMOSHENKO, S. and GOODIER, J. N., *Theory of Elasticity*, McGraw-Hill, New York, 2nd ed., 1951.

Rheology is the science of the deformation and flow of matter. It is concerned with relationships between stresses and strains in materials and their time derivatives. Rheological properties of materials are of interest to many industries such as the paint and oil industries.

Latest trends can be followed by using the *Transactions of the Society of Rheology*, which are published annually by Interscience,

New York. Volume 5 was published in 1962. It contained the majority of papers presented at the Society's 31st Annual Meeting which were mainly concerned with the mechanics of continua and rheology of suspensions. International Congresses on Rheology are held periodically and their proceedings give a good guide to current thinking. The *Proceedings of the Second International Congress on Rheology, Oxford 26–31 July 1953*, Butterworth, London, 1954, has information on the rheological properties of plastics and high polymers, and on the problems of lubrication.

F. R. Eirich's *Rheology*, Academic Press, New York, 1956, is in three volumes. Volume 1 has introductory material and chapters on the various phases of the deformation of solids; Volume 2 includes an integrated survey and discussion relaxation theory, experimental techniques and special types of materials or behaviour; Volume 3 contains chapters on crystalline and cross-linked plastics, inks, pastes, etc., and a survey of industrial rheology.

Societies covering rheology include the *American Society of Rheology*, which is a member society of the American Institute of Physics and was first organized in 1929 with the object of developing the science of the deformation and flow of matter. An annual meeting is held and the society publishes *Rheology Bulletin* three times a year. The corresponding British society is the *British Society of Rheology*, founded in 1940, for whom Pergamon Press publishes the quarterly *Rheology Abstracts*. In this, over 100 periodicals are abstracted regularly. Full abstracts are given of worthwhile articles in obscure or lesser known language journals. The abstracts are arranged in sections covering theory, instruments and techniques, metals and other solids, polymers, elastomers and viscoelastic materials, pastes and suspensions, liquids and general topics.

Friction [Dewey 532 Class]

Friction is the resisting force at the common boundary of two bodies when one of them moves relative to the other. It plays a big part in the operation of clutches, brakes, etc., and in motor-cars where about 20% of the power is wasted overcoming friction. A classic text is by F. P. Bowden and D. Tabor, *The Friction and Lubrication of Solids*, Oxford University Press, London, 1950.

Hydraulics, Fluid Mechanics, Hydrodynamics
[Dewey 532, 532.1, 532.5 Classes]

Hydromechanics is the science of mechanics of fluids and divides into hydrostatics and hydrodynamics. Hydraulics is the branch of hydromechanics which is concerned with the technical aspects of the mechanics of liquids.

Applications of these fields of study are shown in H. Addison's *A Treatise on Applied Hydraulics*, Chapman & Hall, London, 4th ed., 1954, and H. R. Vallentine's *Applied Hydrodynamics*, Butterworth, London, 1959.

V. L. Streeter has edited a useful *Handbook of Fluid Dynamics*, McGraw-Hill, New York, 1961, which covers fluid flow principles, theory, methods and allied data dealing with both fundamental concepts and applied fields.

It is not possible in the space available to discuss all the books on these subjects and the following should be taken as an indicative selection only:

BAYLEY, F. J., *An Introduction to Fluid Dynamics*, Allen & Unwin, London, 1958.

BINDER, R. C., *Fluid Mechanics*, Prentice-Hall, New York, 3rd ed., 1955.

BRENKERT, K., *Elementary Theoretical Fluid Mechanics*, Wiley, New York, 1960.

COLE, G. H. A., *Fluid Dynamics*, Methuen, London, 1962.

DUNCAN, W. *et al.*, *The Mechanics of Fluids*, Arnold, London, 1960.

JONES, J. O., *Introduction to Hydraulics and Fluid Mechanics*, Harper, New York, 1953.

LEWITT, E. H., *Hydraulics and Fluid Mechanics*, Pitman, London, 10th ed., 1958.

RICHARDSON, E. G., *Dynamics of Real Fluids*, Arnold, London, 2nd ed., 1961.

Periodicals include: *Physics of Fluids*, which is published monthly and reports original research in structure dynamics and general physics of gases, liquids and plasmas; and *Journal of Fluid Mechanics*, which covers theoretical and experimental observations and is also published monthly.

Abstracting services include those issued by the British Hydromechanics Research Association as *Abstract Bulletin*. These are compiled by the Research Association and issued bi-monthly.

Research is undertaken by such bodies as the National Physical Laboratory, Teddington, which works on ship dynamics; the British Ship Research Association, London; and the British Hydromechanics Research Association, Harlow, which works on such topics as fundamental fluid mechanics, motion of fluids in pipes, fluid power and oil hydraulics; the National Engineering Laboratory, East Kilbride, is also interested in fluid mechanics including water hydraulic machinery and low speed aerodynamics.

Civil engineering projects involving water from the time it falls on land until the time it reaches the sea are the province of the Hydraulics Research Station, Wallingford. Other research organizations include the Italian Centro Lombardo di Richerche Idrauliche. The international body for hydraulics is the International Association for Hydraulic Research (IAHR). Its function is to promote collaboration between international specialists to enable them to exchange views, experience and knowledge.

Flow Measurement, Turbulence [Dewey 532.5 Class]

Methods of flow measurement depend on the types of fluid involved, rates and quantities. The intruments used all have two parts, one part doing the actual measuring and the other indicating or recording this measurement. The American Society of Mechanical Engineers have surveyed the field in *Fluid Meters, their Theory and Applications*, ASME, New York, 5th ed., 1959, which has extensive bibliographies. Another theoretical and practical review of the present state of the science and art is Linford's *Flow Measurement and Meters*, Spon, London, 2nd ed., 1961.

Turbulence is the motion of fluids with irregular fluctuations in local velocities and pressures. The significant steps in the development of the theory are illustrated by papers published prior to 1950, and collected together by S. K. Friedlander and L. Topper as *Turbulence: Classic Papers on Statistical Theory*, Interscience, New York, 1961. Another well-received work is by J. O. Hinze, *Turbulence*, McGraw-Hill, New York, 1959.

Vacuum Technique [Dewey 533.13 Class]

Vacuum techniques are the methods employed for obtaining a condition of high vacuum. They are particularly applicable in the manufacture of television tubes and electronic valves where they are necessary for the successful operation of the final product.

S. Dusham has reviewed the operations involved in the production and measurement of high and ultra-high vacuum in his *Scientific Foundations of Vacuum Technique*, Wiley, New York, 2nd ed., 1962. Extensive references to the original literature are given. Other books which will repay study include:

HOLLAND-MERTEN, E. L., *Handbuch der Vakuum Technik*, Knappe, Halle Salle, 2nd ed., 1950.

REIMANN, A. C., *Vacuum Techniques*, Chapman & Hall, London, 1952.

Vacuum Physics, Institute of Physics, London, 1951.

Advances in Vacuum Science, Pergamon, London.

The first two volumes of *Advances in Vacuum Science* edited by E. Thomas, contain the proceedings of the First International Congress on Vacuum Techniques. Volume 1 covers fundamental problems in vacuum techniques and ultra-high vacuum; Volume 2 covers vacuum systems and the applications of vacuum in various sciences and techniques. Recent advances in vacuum technology are also covered by the *Ninth National Symposium on Vacuum Technology Transactions*, Pergamon, London, 1963, which contains 94 papers on thin films, space problems and vacuum techniques and components.

The terminology of vacuum technology is covered by the Committee on Standards, American Vacuum Society, which has issued a *Glossary of Terms used in Vacuum Technology*, Pergamon, London, 1958. The British Standard 2951 is a *Glossary of Terms used in Vacuum Technology*, British Standards Institute, London, 1958.

Periodicals include *Vakuum Technik* and *Vacuum*. The latter contains abstracts relating to vacuum science and technology as well as authoritative papers by international experts. The abstracts are grouped under a special system of subject headings.

Societies interested in vacuum science, and from whom information can be obtained, include the American Vacuum Society which promotes communication amongst people interested in vacuum science and technology; the Italian Associazione Italiana de Vuoto, Milan, founded in 1963, and the International Organization for Vacuum Science and Technology (IOVST).

Airfoil Theory, Theory of Gases
[Dewey 533.62, 533.7 Classes]

Aerodynamics is the study of forces acting on bodies in air, and an airfoil is a body whose shape causes it to receive a useful reaction from an airstream moving relative to it. Its relevance in the design of aircraft and lifting devices is easy to see. Books which can be used to study the subject include:

KUETLE, A. M. and SCHETZER, J. D., *Foundations of Aerodynamics*, Wiley, New York, 2nd ed., 1959.

PIERCY, N. A. V., *Complete Course in Elementary Aerodynamics*, English Universities Press, London, 1944.

MILNE-THOMPSON, L. L., *Theoretical Aerodynamics*, Macmillan, London, 3rd ed., 1958.

Societies and research organizations in this field include the British Royal Aeronautical Society, Royal Aircraft Establishment, National Physical Laboratory and the American Institute of Aeronautical Sciences. The relevant specialized abstracting source for this field of physics is *Index Aeronauticus*.

The kinetic theory of gases relates the individual motion of gas particles with the macroscopic properties of a gas. In some senses it may be thought of as part of the more comprehensive study of statistical mechanics. The student of the subject should at some stage consult the works of J. H. Jeans, for example his *Introduction to the Kinetic Theory of Gases*, Macmillan, New York, 1940. Many other texts are valuable, for example:

GUGGENHEIM, E. A., *Elements of Kinetic Theory of Gases*, Pergamon, London, 1960.

KNUDSEN, M., *The Kinetic Theory of Gases: some Modern Aspects*, Methuen, London, 3rd ed., 1952.

PRESENT, R. D., *Kinetic Theory of Gases*, McGraw-Hill, New York, 1958.

Sound [Dewey 534 Class]

Sound is a mechanical disturbance in an elastic medium, that is to say it is an alteration in a property, such as pressure or density, of a medium that can be detected by an instrument or a listener. It is sometimes applied only to disturbances which produce auditory sensations. The various terms in this field are sometimes used in conflicting applications and there is a lack of standard definitions.

Students studying for degrees will find any of the following of value:

CATCHPOOL, E. and SLATTERLEY, J., *Textbook of Sound*, University Tutorial Press, London, 7th ed., 1949.

HUNTER, J. L., *Acoustics*, Prentice-Hall, Englewood Cliffs, New Jersey, 1957.

RICHARDSON, E. G., *Sound. A Physical Textbook*, Arnold, London, 5th ed., 1953.

RSCHEVKIN, S. N., *Theory of Sound*, Pergamon, London, 1962.
WOOD, A. B., *A Textbook of Sound*, Bell, London, 3rd ed., 1955.

Periodicals of value include the cover-to-cover translation of the Russian *Akusticheskii Zhurnal*, which is issued by the American Institute of Physics as *Soviet Physics—Acoustics*. The coverage is mainly physical acoustics, but electro-, bio- and psycho-acoustics and mathematical and experimental research work are also included. A publication of relevance is the *Journal of the Acoustical Society of America*, which is issued monthly and contains original papers on all aspects of acoustics. Other periodicals wholly devoted to the subject are *Acustica and Akustiche Beihefte*.

The Acoustical Society of America's objects are to increase and diffuse the knowledge of acoustics and to prompt its practical application. To fulfil these objects it holds meetings and issues periodicals such as the *Journal* of the Society and *Noise Control*. The British Institute of Physics and the Physical Society has set up a specialist group dealing with acoustics.

Research work is being undertaken on sound waves, velocity and attenuation of sound, by universities and other organizations such as the National Physical Laboratory in Britain, the Acoustics Institute attached to the Department of Physics and Mathematics of the Academy of Sciences of the USSR, the American National Bureau of Standards and industrial laboratories. All these organizations undertake continuing research in the field.

Noise [Dewey 534.35 Class]

Noise is unwanted sound. It has been recognized that exposure to excessively loud noise or prolonged exposure to noise can be very detrimental to health. In the engineering field a change in the type of noise made by machinery is often a warning that something is amiss.

A useful bibliography is *Noise, its measurement, effects and control*, Industrial Hygiene Foundation of America, Pittsburgh, 1955.

The measurement of noise levels is covered by A. P. G. Peterson and E. E. Gross in their *Handbook of Noise Measurement*, General Radio Company, West Concord, Massachusetts, 4th ed., 1960. The National Physical Laboratory has also covered noise measurement in one of its authoritative Laboratory Notes on Applied Science, *Noise Measurement Techniques*, HMSO, London, 1955. (National Physical Laboratory Notes on Applied Science No. 10.)

The various methods applicable to the control of noise are outlined in a handbook edited by C. M. Harris, *Handbook of Noise Control*, McGraw-Hill, New York, 1957. It has forty chapters, written by experts, which give the fundamental practical methods, engineering techniques and reference data on the nature of noise, its measurement and control in buildings, industry, transportation and the community. Ample references are carried by most chapters.

Other useful works include:

Acoustics, Noise and Buildings, Faber & Faber, London, 1958, by P. H. Parkin and H. R. Humphreys; the proceedings of the conferences held by the Acoustics Group of the Institute of Physics and the Physical Society, such as the 1948 *Symposium on Noise and Sound Transmission*, Physical Society, London, 1950, and the proceedings of the conference in 1961 at the National Physical Laboratory, available as *The Control of Noise*, HMSO, London, 1961.

Periodicals covering the measurement and control of noise include *Sound, its Uses and Control*, which is issued bi-monthly and contains practical information on measurement and control of noise, shock, vibration and sound; and *Noise Control*, which is issued by the Acoustical Society of America.

Much research work is being undertaken on such topics as aerodynamic noise, combustion noise, noise in coal mines, noise in urban areas, industrial noise, noise caused by jet aircraft, noise in electrical systems and noise reduction. Research organizations taking part in this research include the Southampton University Institute of Noise and Vibration, and governmental and industrial organizations, such as the Royal Aircraft Establishment, National

Engineering Laboratory, National Physical Laboratory, British Broadcasting Corporation and the Production Engineering Research Association.

Ultrasonics [Dewey 534.5 Class]

Ultrasonics are sound waves with frequencies above the audible. They are finding increasing use in modern science and technology, for example as a method of non-destructive testing, a means of drilling holes in materials, and as a means of cleaning grease and dirt from engineering components, textiles, etc.

Written in simple terms and suitable as an introduction to the science is a book by R. Irving, *Sound and Ultrasonics*, Dobson, London, 1959, whilst a classic text for the serious student is by L. Bergmann, *Der Ultraschall und seine Anwendung in Wissenschaft und Technik*, Hirzel, Stuttgart, 6th ed., 1954. Also well thought of are:

GLICKSTEIN, C., *Basic Ultrasonics*, Rider, New York, 1960.
HUETER, T. F. and BOLT, R. H., *Sonics*, Wiley, New York, 1955.

International conventions on ultrasonics are held periodically and their proceedings are published. They form useful guides to the state of knowledge at a particular time. Research work on the design and use of ultrasonic equipment is undertaken by industry, universities and research institutes such as the Italian Instituto Nazionale di Ultracustica "O. M. Corbino".

Periodicals specifically covering this field include the Iliffe publications *Ultrasonics*.

Questions

1. Choose a book covering rheology and discuss its worth.
2. Consult the books listed under hydraulics and decide which you consider to be the best. Give reasons for your choice.
3. Summarize the latest noise-measuring techniques.
4. What uses have ultrasonics?

CHAPTER 10

Optics

Dewey 535 Class

OPTICS is the study of light. More broadly it covers the study of phenomena associated with the generation, transmission and detection of electromagnetic radiation in the region of the spectrum from about 10 Ångstroms to 1 millimetre. Such recent developments as non-reflecting films and the Schmidt camera testify to the practical importance of its study.

It is convenient to divide optics into three fields: (1) *physical optics*, which deals with the nature of light and is primarily concerned with its wave properties; (2) *geometrical optics*, which treats problems of reflection and refraction from the ray aspects and ignores the wave nature of light; (3) *quantum optics*, which deals with the interaction of light with the atomic entities of matter, and needs quantum mechanics for an exact treatment. Relevant material for this last field will be found in the chapter containing quantum mechanics.

Current progress is reported in *Progress in Optics*, North-Holland, Amsterdam, a series which started in 1961 and which contains review articles about current researches. An excellent survey of recent advances such as the optical image, the Foucault test and the Schmidt camera, is E. H. Linfoot's *Recent Advances in Optics*, Oxford University Press, London, 1955.

The Royal Institution Christmas Lectures often provide a good simple introduction to a scientific subject. Sir William Bragg's lectures in 1931 covering the nature of light, optics, colour, polarization, light from the sun and stars, roentgen rays and the theories of wave and corpuscular nature of light, have been published as *The Universe of Light*, Bell, London, 3rd ed., 1936.

Periodicals

The major optical societies in the world, such as the Optical Society of America, usually publish a scientific journal, e.g. the *Journal of the Optical Society of America*. This is issued monthly and contains original papers on all aspects of optics. The Optical Society of America is also responsible for the publishing of a cover-to-cover translation of the Russian journal *Optikai Spektroskopiya* as *Optics and Spectroscopy* and thus provides an easy way to keep in touch with progress in Russia.

The application of optics is covered by *Applied Optics* which is issued monthly and contains original papers in applied optics and related fields. Optics and human vision is the field of two periodicals published by the British Optical Association, the quarterly *British Journal of Physiological Optics* and the fortnightly *Ophthalmic Optician*.

Societies and Research Organizations

Besides publishing periodicals on optics and human vision, the British Optical Association has an interesting collection of antique optical instruments and specimens illustrating the history of spectacles. The international body is the International Commission for Optics which holds periodic meetings such as the one held in Munich in September 1962.

The Optical Society of America was founded in 1916 and aims at increasing and diffusing the knowledge of optics and encouraging co-operation between all interested parties. Conferences are held, periodicals published and medals awarded for distinguished work in optics. The Institute of Physics and the Physical Society have set up a specialist subject group dealing with optics and specialist meetings are held regularly.

Research work on optics in Britain is undertaken by the National Physical Laboratory, universities, industrial laboratories, etc. The National Physical Laboratory has, amongst its several divisions, a Light Division which has three sections dealing with the study of radiation in the ultraviolet, visible and infrared regions of the spectrum, photometry and colorimetry including research in vision, and optics. Recent research covers the development of a spectro-polarimeter, methods of increasing accuracy in engineering metrology, interferometry for measuring the flatness of optical surfaces and production of reflecting surfaces for solid state optical masers.

European research organizations include the Institut D'Optique Théorique et Appliquée in Paris and the Instituto Nazionale di Ottica in Firenze, Italy.

Physical Optics [Dewey 535.2 Class]

Physical optics, that part of optics involving the wave properties of light, is well represented by texts suitable for university students such as C. Curry's *Wave Optics: Interference and Diffraction*, Arnold, London, 1957, and the very well used work by F. A. Jenkins and H. E. White, *Fundamentals of Optics*, McGraw-Hill, New York, 3rd ed., 1957, whilst R. A. Houston's *Physical Optics*, Blackie, London, 1957, was written for students who have reached the intermediate science standard in physics.

Recent research in diffraction of transverse waves in the near as well as far regions together with general methods of solving problems of wave optics is covered by C. L. Andrews in his *Optics of the Spectro-magnetic Spectrum*, Prentice-Hall, Englewood Cliffs, New Jersey, 1960. Other good texts include:

BORN, M. and WOLF. E., *Principles of Optics*, Pergamon, London, 1959.
DITCHBURN, R. W., *Light*, Blackie, London, 2nd ed., 1963.
SHURCLIFF, W. A., *Polarized Light*, Harvard University Press, Cambridge, Massachusetts, 1962.

Photometry [Dewey 535.22 Class]

Photometry is that aspect of optics which is concerned with the measurement of the luminous intensities of light sources and of luminous flux and illumination and is the subject of one of the classic treatises of science by P. Bouger, *Optical Treatise on the*

Gradation of Light, University of Toronto Press, Toronto, 1961. The original was published in 1760 and the author can be considered the inventor of photometry. A more recent publication which describes both theoretical and practical matters affecting the measurement of light flux, candle power, illumination, etc., but excludes any treatment of the use to which such measurements can be put, is:

WALSH, J. W. T., *Photometry*, Constable, London, 3rd ed., 1958.

Geometrical Optics [Dewey 535.3 Class]

Geometrical optics is the most useful level at which to work when designing optical instruments because it is that part of optics where the size of the wavelength can be neglected.

A useful bibliography is S. S. Ballard's *Bibliography on Reflecting Optics Covering the Period 1925–1950*, University of Michigan, Ann Arbor, 1950.

Useful texts include:

COX, A., *Optics, the Technique of Definition*, Focal Press, New York, 10th ed., 1953.
CURRY, C., *Geometrical Optics*, Arnold, London, 1953.
LONGHURST, R. S., *Geometrical and Physical Optics*, Longmans, London, 1957.
MACKINNON, L., *A Textbook on Light*, Longmans, London, 1961.
MARTIN, L. C., *Geometrical Optics*, Pitman, London, 1955.
PITCHFORD, A., *Studies in Geometrical Optics*, Macdonald, London, 1959.
SMITH, C. J., *Optics*, Arnold, London, 1960.

Thin films have been the subject of much study during the last few years, by the National Physical Laboratory amongst others, and this study is continuing. The uses of thin films include that of multilayer filters and as a protective on the hygroscopic optical components used in infrared spectroscopy. Because of this continuing interest, A. Vasicek's *Optics of Thin Films*, North-

Holland, Amsterdam, 1960, is very useful. It is a survey and history of thin films in optics which have developed in the last two decades.

Of prime importance to the optical instrument designer is a book by R. J. Bracey, *The Technique of Optical Instrument Design*, English Universities Press, London, 1960. Some indication of its worth can perhaps be gained from its chapter headings, which are: introductory survey; the use of ray traces; wave theory and the general laws covering asymmetrical abbreviations; analysis and control of axial aberrations; design of eyepieces; telescopes, periscopes and the microscope; advanced studies in optical design; image assessment; illumination and lens systems.

Luminescence, Fluorescence and Phosphorescence
[Dewey 535.35 Class]

Luminescence is the emission of light by a body where such emission cannot be attributed merely to the temperature. Fluorescence and phosphorescence are special cases of luminescence, the study of which assists in the detection and measurement of high energy radiations. The subject can be studied by using H. W. Leverenz's *An Introduction to Luminescence of Solids*, Wiley, New York, 1950, and P. Pringsheim's *Fluorescence and Phosphorescence*, Interscience, New York, 1949.

An important meeting between transistor physics and the disciplines developed in connection with photoluminescence is electroluminescence. H. K. Hensich's *Electroluminescence*, Pergamon, London, 1962, includes a bibliography and gives a general survey, with a theoretical treatment, and experiments on single crystal phosphors and microcrystalline phosphors.

Colour [Dewey 535.6 Class]

Colour is one aspect of optics that everyone is conscious of. The usefulness of a scientific study of the subject is shown when one considers the importance of regularity of colour to paint and ink manufacturers. Various methods of measuring colour are available, some depending on the human eye and others on more sophisticated means such as spectroscopy.

Simple introductions are:

EVANS, R. M., *An Introduction to Color*, Wiley, New York, 1948.

HARTRIDGE, H., *Colours and How We See Them*, Bell, London, 1949.

Textbooks include:

BOUMA, P. J., *Physical Aspects of Colour*, Philips Gloeilampen-fabricken, Eindhoven, 1948.

MURREY, H. D. (ed.), *Colour in Theory and Practice*, Chapman & Hall, London, 1952.

The Science of Colour, Cromwell, New York, 1953.

Principles, methods and applications of the trichromatic systems of colour measurement are outlined by W. D. Wright in *The Measurement of Colour*, Hilger & Watts, London, 2nd ed., 1958, whilst the Ostwald system is covered in E. Jacobson's *Basic Colour: an Interpretation of the Ostwald Colour System*, Theobald, Chicago, 1948.

Spectroscopy [Dewey 535.842 Class]

The original range of spectroscopy was limited to observing spectra formed by dispersing light, but now the term has been extended to cover electromagnetic radiations—infrared, ultraviolet, X-rays, gamma-rays, visible and microwave. The use of spectroscopy as an analytical tool, as in the spectrochemical analysis of metals by means of the line spectra they are made to emit, is extensively used in laboratories throughout the world.

Current research work is reviewed in the annual *Advances in Spectroscopy*, Interscience, London. Volume 2, published in 1962, covered the application of atomic absorption spectra to chemical analysis, spectra of flames, X-ray spectroscopy, nuclear magnetic resonance, infrared spectra of crystals, refraction of gases in the infrared, infrared spectroscopy of micro-organisms, ultraviolet absorption spectra of proteins and related compounds and some recent developments in the theory of molecular energy levels.

Proceedings of the various scientific meetings also usually indicate the present state of the art and are much valued contributions to the literature. *Spectroscopy*, Pergamon, London, 1962, is the proceedings, edited by M. J. Wells, of a conference organized by the Institute of Petroleum in March 1962, and which presents a balanced picture of spectroscopy at the time. Others available include the *Proceedings of the Sixth International Spectroscopy Colloquim*, Pergamon, London, 1957, the proceedings of the 4th Biennial Meeting of the European Spectroscopy Group which are contained in *Advances in Molecular Spectroscopy*, Pergamon, London, 1962, and the proceedings of the 5th European Congress on Molecular Spectroscopy published as *Molecular Spectroscopy*, Butterworth, London, 1962. G. L. Clark has edited a most valuable encyclopaedia, *The Encyclopaedia of Spectroscopy*, Reinhold, New York, 1960. This is an authoritative survey of the entire field of spectroscopy, starting with absorption spectrophotometry and running alphabetically through X-ray emission spectra. Over 160 internationally recognized contributors are responsible for the articles, the majority of which have references for further reading.

Indexes and Tables

Spectroscopy is one field of physics in which much data has been collected in tabular form and is readily available for consultation, thus saving time which would be spent in seeking the information from various diverse sources. To give a general indication of the sort of tables available, a few of them are listed:

AHRENS, L. H., *Wavelength Tables of Sensitive Line*, Addison-Wesley, Cambridge, Massachusetts, 1951.

HARRISON, G. R., *Wavelength Tables with Intensities in Arc, Spark or Discharge Tube of More Than 100,000 Spectrum Lines*, Wiley, New York, 1939. A most important collection.

KAYSER, H., *Tabelle der Schwingungszahlen der auf das Vakuum*

OPTICS

reduzierten Wellenlangen zwischen 2,000 Å and 10,000 Å,
Hirzel, Leipzig, 1925.

KAYSER, H. and RITSCHL, R., *Tabelle der Hauptlinien der Linien-*
spektren aller Elemente nach Wellenlangen geordnet, Springer,
Berlin, 2nd ed., 1939.

MOORE, C. E., *An Ultraviolet Multiplet Table: The Spectra of*
Chromium, Manganese, Iron, Zirconium and Niobium, GPO,
Washington D.C., 1952. (U.S. National Bureau of Standards
Circular 488 Section 2.)

MOORE, C. E., *An Ultraviolet Multiplet Table: The Spectra of*
Hydrogen, Lithium, Titanium and Vanadium, GPO, Washington
D.C., 1950. (U.S. National Bureau of Standards Circular 488
Section 1.)

H. M. Hershenson has compiled indexes to the literature for the
period 1930–59.

HERSHENSON, H. M., *Ultraviolet and Visible Absorption Spectra:*
Index for 1930–1954 and 1955–1959, Academic Press, New York,
1956, 1961.

Periodicals

Several periodicals are available and can be used to keep reasonably up to date with the latest developments in spectroscopy.
They include the bi-monthly *Applied Spectroscopy* which covers
the theory and practice of absorption spectroscopy—X-ray, ultraviolet, infrared, visible and microwave; emission spectroscopy—
arc and spark, flame and fluorescence; Raman spectroscopy,
diffraction, mass spectroscopy and nuclear magnetic resonance
spectroscopy.

Covering the application of optical spectroscopy to chemical
problems is the monthly *Spectrochimica Acta* which contains
original papers dealing with emission and absorption spectroscopy
over the entire optical and X-ray wavelength range, and spectroscopy in the microwave region in so far as it has a direct physico-
chemical interest. Reports of meetings, review articles and book
reviews are included. Russian work can be examined using the
cover-to-cover translation, for which the Optical Society of

America are responsible, of the Russian *Optikai Spektroskopiya*, published in English as *Optics and Spectroscopy*.

Abstracts

Spectroscopy is covered by the general abstracting publications, which have been detailed in a separate chapter, but as well as these there are several specialized publications. *Spectrographic Abstracts* issued by the Technical Information and Library Services of the Ministry of Aviation, London, is one of these. Infrared and Raman spectroscopy are covered and the publication is available from Her Majesty's Stationery Office. Publication is infrequent and a few years behind the current literature as instanced by the publication in 1961 of the abstracts covering 1957. Despite this, it is, of course, still very valuable. Name, formula and author indexes are included.

Hilger & Watts used to publish *Spectrochemical Abstracts* which covered emission spectroscopy in its analytical applications and also flame spectroscopy but not X-ray emission spectroscopy. However, publication ceased with Volume VIII which covered the period 1958–61.

Index to the Literature on Spectrochemical Analysis is issued periodically by the American Society for Testing Materials; Part 1 covers the period 1920–39, Part 2 1940–45, Part 3 1946–50, Part 4 1951–55. References with abstracts are listed chronologically by year and in alphabetical order of the first author's name. A detailed subject and author index is included.

The American Society for Applied Optics has a section of its *Spectroscopy Bulletin* devoted to the *Literature of Applied Spectroscopy*, both U.S. and foreign publications being covered.

Societies and Research Organizations

Societies specifically for the spectroscopist include the American Society for Applied Spectroscopy which was formed in 1945 with the purpose of advancing the science and practice of applied spectroscopy. As well as holding meetings, a medal is awarded annually for an outstanding contribution to spectroscopy, and the society publishes *Applied Spectroscopy*. In Britain, the Institute of Physics and the Physical Society have set up a specialist subject group dealing with applied spectroscopy.

Research is being undertaken in many countries and by many organizations, industrial, governmental and universities. Germany has the Institut für Spektrochemie und angewandte Spektroskopie at Dortmund-Aplerbeck which has done work on the physics of spectrochemical light sources and spectrochemical investigations of ores, soils, etc., and there is also the Forschungsstelle für Spektroskopie in der Max-Planck-Gesellschaft zFdW Hechingen/ Hohenzollern. In the United States, the National Bureau of Standards undertakes work on spectrochemistry, infrared, far ultraviolet, plasma and molecular spectroscopy in some of its many divisions, whilst in Britain the National Physical Laboratory does research work covering spectroscopy. Other organizations throughout the world can easily be located from the various guides and they form a most useful source of information.

Books

Students of spectroscopy have many textbooks, such as the standard comprehensive work by E. U. Condon and G. H. Shortley, *The Theory of Atomic Spectra*, Cambridge University Press, London, 1959, from which to choose. The following should be taken as a small selection from amongst the many books of value:

BARROW, G. M., *Introduction to Molecular Spectroscopy*, McGraw-Hill, New York, 1962.

BOHR, N., *Theory of Spectra and Atomic Construction*, Cambridge University Press, London, 1962.

BREENE, R. G., *The Shift and Shape of Spectral Lines*, Pergamon, London, 1961.

KUHN, H. G., *Atomic Spectra*, Longmans, London, 1962.

MELLON, M. G. (ed.), *Analytical Absorption Spectroscopy: Absorptimetry and Colorimetry*, Wiley, New York, 1950.

PENNER, S. S., *Quantitative Molecular Spectroscopy and Gas Emissivities*, Pergamon, London, 1959.

WALKER, S. and SHAW, H., *Spectroscopy: Vol. 1, Microwave and Radio Frequency Spectroscopy*, Chapman & Hall, London, 1961.

WILSON, E. B. *et al.*, *Molecular Vibrations: the Theory of Infrared and Raman Vibrational Spectra*, McGraw-Hill, New York, 1955.

A valuable examination of the practical aspects of spectroscopy is V. S. Burakov's and A. A. Yankovskii's *Practical Handbook on Spectral Analysis*, Pergamon, London, 1963. The major practical problems of visual and photographic methods of spectral analysis are dealt with and the book contains atlases of spectra and tables of spectral lines.

Infrared Spectra [Dewey 535.842 Class]

Indexes and Tables

Valuable compilations have been made of data on infrared spectra such as *An Index to Published Infrared Spectra*, HMSO, London, 1960, which covers most of the spectra published up to 1957. An index by formula is included. Others equally valued include H. M. Hershenson's *Infrared Absorption Spectra Index for 1947–57*, Academic Press, New York, 1959, which gives about 16,000 references to published information in 33 American and European journals and one book, with an index arranged according to the compounds whose spectra is given; and the National Bureau of Standards *Infrared Bibliography*, Washington D.C., 1952–. Calibration of spectrometers is aided by *Tables of Wavenumbers for the Calibration of Infrared Spectrometers*, Butterworth, London, 1961.

Society

The Coblentz Society was established in the U.S.A. in 1955 to foster understanding and application of infrared spectroscopy. It publishes infrared analytical methods in *Analytical Chemistry*.

Books

Those interested in studying this subject could use the translated version of the 3rd edition of W. Brugel's *An Introduction to*

Infrared Spectroscopy, which was published by Methuen, London, in 1962. As well as basic theory, instrumentation, applications, frequency correlations of some selected groups and the spectra of high polymers are covered. G. K. T. Conn and D. G. Avery are responsible for the very readable *Infrared Methods: Principles and Applications*, Academic Press, New York, 1960, which deals with sources, optical materials, filters, detectors, amplifiers, dispersive systems and practical applications. P. W. Kinse's *Elements of Infrared Technology*, Wiley, New York, 1962, is also very useful.

Questions

1. What are the latest developments in optics?
2. Compare and contrast the various abstracts covering spectroscopy.
3. What is the Ostwald colour system?

CHAPTER II

Heat

Dewey 536 Class

HEAT is a form of energy associated with the motion of atoms or molecules of a body and the subject encompasses temperature, thermometry, heat conduction, thermodynamics, cryogenics, etc. The science is applied in such fields as temperature measurement and heating and ventilating.

Bibliography

A good bibliography is prepared by the National Engineering Laboratory, East Kilbride, Scotland—*Heat Bibliography*, HMSO, London. Abstracts are grouped under subject headings such as boiling, combustion, condensation, conduction, convection, evaporation, fins and extended surfaces, fluid flow, freezing and melting, heat engines, heat transfer coefficients and media, insulation, high temperature, physical properties, radiation thermodynamics. Editions have been published covering the period 1948–59.

Research Organizations

Many of the countries of the world have governmental laboratories which deal in some way with heat. The National Bureau of Standards has a division for heat, the sections of which concern themselves with temperature physics, heat measurements, cryogenic physics, equation of state and statistical physics.

Books

An introductory course in heat which requires no previous knowledge is T. Samson's *Mechanics and Heat*, Macdonald & Evans, London, 1960. Taking the subject up to the requirements of the ordinary level of the General Certificate of Education is *Heat and Light*, English Universities Press, London, 1959, by R. Stone and others.

The number of editions published shows the value of M. W. Zemansky's *Heat and Thermodynamics*, McGraw-Hill, New York, 4th ed., 1957. This is an intermediate text for students of physics, chemistry and engineering. J. K. Roberts and A. R. Miller have written a classic text for honours degree students, *Heat and Thermodynamics*, Blackie, London, 5th ed., 1960.

These four publications are just a few of the many which will repay study, others can be found from library catalogues or published book lists.

Heat Transfer [Dewey 536.2 Class]

Heat may be transmitted by conduction through a medium, by convection currents in a fluid, or by radiation which can pass through empty space. Industry is interested in heat transfer when designing heat exchangers, which extract heat from one fluid stream and add it to another, and when specifying the fluid to be used in such heat exchangers. Furnace designers, heating engineers and others are all concerned with heat transfer.

Recent developments can be found from *Heat Transfer and Fluid Mechanics Institute*, Stanford University Press, Stanford, California, which is issued yearly and reprints papers from the annual meeting covering scientific and technical work of a fundamental nature; and from *Recent Advances in Heat and Mass Transfer*, McGraw-Hill, New York. The proceedings of other conferences which can be studied with value include *International Developments in Heat Transfer*, American Society of Mechanical Engineers, New York, 1961, which is the proceedings of the 1961 International Heat Transfer Conference.

Periodicals

Recent advances and new ideas in heat transfer can be followed by reading periodicals such as the *International Journal of Heat and Mass Transfer*, which provides a medium for the exchange of basic ideas in heat and mass transfer between research workers and engineers located throughout the world. The emphasis is on original research, both analytical and experimental. Special review articles on new advances appear from time to time. An international bibliography of recent papers in heat and mass transfer listed by title, author and source appears in each issue.

The American Society of Mechanical Engineers issues quarterly the *Journal of Heat Transfer* which contains original papers which are not published elsewhere. Abstracts of articles on heat transfer appear in *Engineering Index* and in *Fuel Abstracts*. The former is an alphabetical index whilst in the latter abstracts appear in the monthly issues under broad subject headings. An annual subject index is issued.

Societies and Research Organizations

All of the societies throughout the world concerned with mechanical engineering such as the Institution of Mechanical Engineers and the American Society of Mechanical Engineers have an interest in heat transfer. The American Society of Mechanical Engineers which was first organized in 1880 has a special division concerned with heat transfer. The purpose of the society is to promote the art and science of mechanical engineering and the allied arts and sciences; to encourage original research; to foster education and advance the standards of engineering; to promote intercourse of engineers amongst themselves and with allied technologists. The societies concerned with heating and ventilating, such as the Institution of Heating and Ventilating Engineers in Britain, also have an interest in heat transfer. They hold meetings and issue publications and in this way provide information on new techniques.

The Heat Transfer and Fluid Mechanics Institute of Stanford, California, exists to present technical and scientific advances in heat transfer and fluid mechanics. It does this by doing research and holding conferences.

Amongst the governmental laboratories dealing with heat

transfer and heat exchange apparatus are the National Engineering Laboratory, East Kilbride, and those set up by the various bodies dealing with coal, gas, electricity and atomic power, the National Coal Board, the Gas Council, the Central Electricity Generating Board and the United Kingdom Atomic Energy Authority.

Other research organizations include the Austrian Bundesversuchs- und Forschungsanstalt für Wärme-Kälte- und Strömungstechnik (WKS), Arsenal Objekt 222, Wien III, which is an experimental and research station for government, industry and trade in the field of heat, cold and flow technology, and the British Heating and Ventilating Research Association, Bracknell, which concerns itself with topics such as the heating and ventilating of large buildings and design procedures for high temperature radiant heating.

Books

Many textbooks deal with heat transfer, a selection of which follows:

CHAPMAN, J. H., *Heat Transfer*, Macmillan, New York, 1960.

ECKERT, E. R. G. and DRAKE, R. M., *Heat and Mass Transfer*, McGraw-Hill, New York, 2nd ed., 1959.

FISHENDEN, M. W. and SAUNDERS, O. A., *An Introduction to Heat Transfer*, Clarendon Press, Oxford, 1950.

JAKOB, M. and HAWKINS, G. A., *Elements of Heat Transfer*, Wiley, New York, 3rd ed., 1957.

KAY, J. M., *An Introduction to Fluid Mechanics and Heat Transfer*, Cambridge University Press, London, 1957.

KNUDSEN, J. G. and KATZ, D. L., *Fluid Dynamics and Heat Transfer*, McGraw-Hill, New York, 1958.

LYKOV, A. V. and MIKHAYLOV, Y. A., *Theory of Energy and Mass Transfer*, Prentice-Hall, Englewood Cliffs, New Jersey, 1961.

ROHSENOW, W. M. and CHOI, H. Y., *Heat, Mass and Momentum Transfer*, Prentice-Hall, London, 1961.

SMALL, J., *First Steps in Heat Transfer*, Blackie, London, 1959.

TYRELL, H. J. V., *Diffusion and Heat Flow in Liquids*, Butterworth, London, 1961.

WRANGHAM, D. A., *The Elements of Heat Transfer*, Chatto & Windus, London, 1961.

One of the recognized experts is S. S. Kutateladze and he has collaborated with V. M. Borishanskii in the production of *A Concise Encyclopaedia of Heat Transfer*, Pergamon, London, 1965, which is a collection of data and formulae used in calculations on all types of heat transfer problems covering basic principles of the theory of heat exchange, heat conductivity, convection, radiation, principles of the thermal and hydraulic calculation of instruments of heat exchange. A bibliography is included.

Thermometry and Pyrometry
[Dewey 536.51, 536.52 Classes]

Thermometry is concerned with the measurement of the relative degree of hotness of a substance. Various methods are available, liquid-filled thermometers, gas thermometers, electric thermometers—thermocouples, electric resistance thermometers, magnetic susceptibility thermometers, optical thermometers—pyrometers, and strain thermometers. The measurement of temperature has many practical applications, particularly as a process control tool in industry.

Books

Papers presented at symposia held in 1939, 1954 and 1961 have been collected in the three-volume *Temperature: its Measurement and Control in Science and Industry*, Reinhold, New York. Basic concepts, standards and methods, and applied methods and instruments are covered and these three volumes should be required reading for anyone interested in a serious study of the subject.

Other texts include:

Calibration of Temperature Measuring Instruments, HMSO, London, 2nd ed., 1957.

CAMPBELL, C. H., *Modern Pyrometry*, Chemical Publishing, New York, 1951.

HALL, J. A., *Fundamentals of Thermometry*, Institute of Physics, London, 1953.

HALL, J. A., *Practical Thermometry*, Institute of Physics, London, 1953.

HARRISON, T. R., *Radiation Pyrometry and its Underlying Principles of Radiant Heat Transfer*, Wiley, New York, 1960.

A symposium was held in Chicago in March 1960 the objects of which were (1) to promote a fuller understanding of the meaning of temperature and the plasma state of matter, (2) to provide a discussion forum to explore the validity of the theoretical premises and to provide a critical review of experimental techniques, (3) to highlight the areas worthy of further study. The proceedings of the symposium have been edited by P. J. Dickerman and published as *Optical Spectrometric Measurements of High Temperatures* and they give a good indication of thinking current at the time.

Periodicals and Abstracts

Amongst the periodicals which include material on thermometry, and which enable one to keep abreast of the latest research and developments, is the *Journal of Scientific Instruments*. This is published monthly by the Institute of Physics and the Physical Society in association with the National Physical Laboratory and it contains articles describing physical instruments and instrumental and general experimental techniques. Other periodicals covering instruments which include articles on thermometry are *Instrument Practice*, *Instrument Society of America Journal*, *Instrument Engineer*, which is published twice a year and also carries selected references, *Instrument Construction*, which is a translation of the Russian monthly *Priborostroenie* and is produced by the British Scientific Instrument Research Association, *Transactions of the Society of Instrument Technology* and the *Review of Scientific Instruments*.

References to articles on thermometry are included in the section on heat in *Instrument Abstracts*. The abstracts are compiled by the British Scientific Instrument Research Association and published monthly by Taylor & Francis. Each issue has an author index and annual author and subject indexes are published.

Research Organizations and Societies

The British National Physical Laboratory maintains and improves the British primary standards of temperature. Recent work includes a comparison of the temperature scales (10–90°K) of the national laboratories of U.S.A., U.S.S.R. and the U.K., and work on the establishment of the thermodynamic temperature scale between 175 and 1063°C by means of Planck's law, using monochromatic infrared radiation. The laboratory is represented on the Advisory Committee for Thermometry of the International Committee for Weights and Measures, also on the various British Standards Institute committees which deal with resistance thermometers and thermocouples, and on the International Standards Organization working group on thermometers. The comparable American organization is the National Bureau of Standards.

Other research organizations which have some interest in thermometry include the British Scientific Instrument Research Association, whose work includes fundamental studies of all techniques relating to scientific and industrial instruments, the evaluation of instruments and the development and construction of novel and special instruments. The Technisch Physische Dienst TNO en TH, Delft, has advisory services and undertakes physical research which includes work on heat measurement and thermometry. Then there are the research facilities in industrial firms and these can be found from the various directories.

The Society of Instrument Technology was founded in Britain in 1945 with the object of promoting the development and progress of instrumentation technology, especially that applied to industrial manufacturing, transport, communication and defence. Meetings are held both nationally and by local branches, The more important papers read at these meetings are printed in the quarterly *Transactions of the Society of Instrument Technology*. The American equivalent society is the Instrument Society of America which was first organized in 1939, though under a different name, changing to its present title in 1946. The Society's purpose is to advance the arts and sciences connected with theory, design, manufacture and use of instruments in the various sciences and technologies. Various committees such as that on physical properties measurement meet regularly and the monthly *Instrument Society of America Journal* is published.

Low Temperature Physics, Cryogenics
[Dewey 536.56 Class]

This is the general field of scientific work at low temperature. Work on low temperatures has found many applications such as the storage and handling of liquid hydrogen for space applications, oxygen breathing apparatus for high altitude flights, solid state memory and amplifier devices and isotope separation.

To follow the latest developments "progress reports" may be used such as: MENDELSSOHN, K. (ed.), *Progress in Cryogenics*, Academic Press, New York, and GORTER, C. J. (ed.), *Progress in Low Temperature Physics*, North-Holland, Amsterdam. Volume 3 was published in 1961 and covered vortex lines in liquid helium II, helium ions in liquid helium II, the nature of the λ transition in liquid helium, liquid and solid helium, ^{3}He cryostats, recent developments in super conductivity, electron resonances in metals, orientation of atomic nuclei at low temperatures, solid state masers, the equation of state and the transport properties of the hydrogenic molecules, some solid gas equilibria at low temperatures.

Periodicals and Bibliographies

Specialized periodicals include *Cryogenics* which is issued quarterly and contains original papers and technical notes on all aspects of low temperature engineering, research and developments.

Abstracts of articles on low temperature physics appear in those abstracting journals which cover all fields of physics. A list of current literature also appears in *Cryogenics*. A bibliography covering the period 1944–60 was issued as a supplement to this periodical with the title *A Bibliography of Low Temperature Engineering and Research, 1944–1960*, and this is a most valuable source of references.

Societies and Research Organizations

A special subject group of the Institute of Physics and the Physical Society has been set up to cover low temperature physics and certainly the major physical society of a particular country will be interested in this topic. The International Institute of Refrigeration, Paris, has an interest in cryogenics as shown by the papers read at some of its meetings. The Institute was set up in 1920 to study all technical, scientific, economic and industrial questions concerning refrigeration. A technical board directs International Commissions as to the research work they should undertake.

As well as many industrial and university laboratories working in the field of cryogenics some governments have also set up special laboratories. One of these is the Cryogenic Engineering Laboratory of the American National Bureau of Standards who are working on cryogenic processes, cryogenic properties of solids and properties of cryogenic fluids. The heat division of the Bureau also has an interest in cryogenic physics.

Books

Once again there is a considerable number of books which deal with low temperature physics and those mentioned in the following paragraphs should be taken as no more than an indicative selection.

The student should be aware of the lectures given at the Royal Institution by F. E. Simon, N. Kurti, J. F. Allen and K. Mendelssohn which have been reprinted in their original form, with some revisions to include more recent materials, as *Low Temperature*

Physics, Pergamon, London, 1952. These give a very clear picture of knowledge at that time.

Several conferences have concerned themselves with low temperature physics and the proceedings of most of these conferences have been published and are thus available for consultation. The 7th International Conference was held at the University of Toronto in 1960 and the proceedings are available as *Low Temperature Physics*, North-Holland, Amsterdam, 1961. Commission I of the International Institute of Refrigeration have held meetings at Delft, Eindhoven and London during the last few years. The meetings covered cryogenic apparatus, thermometry, transport phenomena in liquids and gases, etc., and have been published as a three-volume set edited by A. Van Itterbeck, *Problems of Low Temperature Physics and Thermodynamics*, Pergamon, London, 1959, 1962, 1963. The National Bureau of Standards held a semi-centennial symposium on low temperature physics in 1951 and the papers were published as *Low Temperature Physics*, U.S. Government Printing Office, Washington D.C., 1952.

Other useful titles include:

ATKINS, K. R., *Liquid Helium*, Oxford University Press, London, 1959.
DIN, F. and COCKETT, A. H., *Low Temperature Techniques*, Newnes, London, 1960.
JACKSON, L. C., *Low Temperature Physics*, Methuen, London, 5th ed., 1962.
LANE, C. T., *Superfluids Physics*, McGraw-Hill, New York, 1962.
MENDELSSOHN, K., *Cryophysics*, Interscience, New York, 1960.

Anyone requiring details of experimental techniques is well catered for by, amongst others, a book by G. K. White, *Experimental Techniques in Low Temperature Physics*, Oxford University Press, London, 1959, which covers production and measurement of low temperatures, handling of liquefied gases on a laboratory scale, principles and some details of the design of experimental cryostats, physical data of solids used in making the equipment; and one edited by F. E. Hoare, *Experimental Cryo-*

E

genics, Butterworth, London, 1961, which contains review articles
on low temperature laboratories, the mathematics of gas lique-
faction and liquefier design, liquid air production, the production
of liquid hydrogen and helium, ancillary equipment for the production of liquid hydrogen and helium, materials and methods for
the construction of low temperature apparatus, storage and transfer of liquefied gases, magnetic cooling, low temperature thermometry, cryogenic techniques and miscellaneous applications.

Thermodynamics [Dewey 536.7 Class]

Thermodynamics is concerned with the conversion of energy to
and from heat and the methods employed for such transformation. The science can be said to date from a paper on the subject
given by Clausius in 1850. The chief applications are with respect
to heat engines and to chemical reactions.

Many industrial laboratories such as those of the oil and motor-
car companies do work on the thermodynamics of combustion,
and university departments of chemistry, mechanics and applied
mathematics all are concerned with this subject. The current
listing of research work in British universities gives details of
twenty-seven laboratories at which research is being undertaken.
The National Chemical Laboratory has a Chemical Thermodynamics Group which measures the physico-chemical constants
of pure compounds. European laboratories interested in thermodynamics include the Belgian Institut Belge des Hautes Pressions,
which works on the thermodynamic properties of gases, and the
Austrian Institut für Verfahrenstechnik und Technologie der
Brennstoffe der Technischen Hochschule Wien who investigate
the thermodynamics and kinetics of different gas reactions.

Tables and Bibliography

Many tables of thermodynamic properties are available from
which to locate required information. These include:

ELLENWOOD, F. O. and MACKEY, O. O., *Thermodynamic Charts:
Steam, Water, Ammonia, Freon 12 and Mixtures of Air and*

Water Vapours, also Special Tables for Turbine Calculations,
Wiley, New York, 1944.

HILSENRATH, J. *et al.*, *Tables of Thermodynamic and Transport Properties of Air, Argon, Carbon Dioxide, Carbon Monoxide, Hydrogen, Nitrogen, Oxygen and Steam*, Pergamon, London, 1960.

KEENAN, J. H. and KAYE, J., *Gas Tables: Thermodynamic Properties of Air, Products of Combustion and Component Gases, Compressible Flow Functions*, Wiley, New York, 1948.

KEENAN, J. H. and KEYES, F. G., *Thermodynamic Properties of Steam, Including Data for the Liquid and Solid Phases*, Wiley, New York, 1936.

PREDVODITELEV, A. S., *Tables of Thermodynamic Functions of Air for the Temperature Range 6000–12,000° and Pressure Range 0·001–1,000 atm.*, Infosearch, London, 1958. (Translated from the 1957 Russian edition.)

STULL, D. R. and SINKE, G. C., *Thermodynamic Properties of the Elements*, American Chemical Society, Washington D.C., 1956. This covers heat capacity, heat content, entropy and free energy, function of the solid, liquid and gas states of the first ninety-two elements.

Thermophysical properties of materials which melt at approximately 1000°F, which covers elements, alloys, ceramics, cermets, intermetallics, composites and polymerics, are brought together in the *Handbook of Thermophysical Properties of Solid Materials*, Pergamon, London, 1961, 1963. This work, which was edited by A. Goldsmith and others, is issued in loose-leaf form so that inclusion of new data or substitution of old is possible. Each of the four volumes of data consists of four sections: (1) introductory remarks and explanatory text, (2) materials index, (3) tables of conversion factors, (4) body of data. The fifth volume has a materials index, an author index and a list of references.

The major abstracting periodicals in physics cover thermodynamics. An early bibliography is available for consultation:

E*

122   HOW TO FIND OUT ABOUT PHYSICS

TUCKERMAN, A., *Index to the Literature of Thermodynamics*, Smithsonian Institution, Washington D.C., 1890. (Smithsonian Miscellaneous Collections, Volume 34, No. 741.)

Books

As has been pointed out previously the value of particular books depends on the user, so the following books should be taken only as an indicative selection:

ASTON, J. G. and FRITZ, J. J., *Thermodynamics and Statistical Thermodynamics*, Wiley, New York, 1959.

CALLEN, H. B., *Thermodynamics: An Introduction to the Physical Theories of Equilibrium Thermostatics and Irreversible Thermodynamics*, Wiley, New York, 1960.

DURHAM, F. P., *Thermodynamics*, Prentice-Hall, Englewood Cliffs, New Jersey, 2nd ed., 1959.

FAIRES, V. M., *Thermodynamics*, Macmillan, New York, 3rd ed., 1957. And an extraction from *Elementary Thermodynamics*, Macmillan, New York, 3rd ed., 1957.

DE GROOT, S. R. and MAZAR, P., *Non-Equilibrium Thermodynamics*, North-Holland, Amsterdam, 1962.

LANDSBERG, P. T., *Thermodynamics: with Quantum Statistical Illustrations*, Interscience, New York, 1961.

LEWIS, G. N. and RANDALL, M., *Thermodynamics*, McGraw-Hill, New York, 2nd ed., 1961.

MARTER, D. H., *Thermodynamics and the Heat Engine*, Thames & Hudson, London, 1960.

OBERT, E. F., *Concepts of Thermodynamics*, McGraw-Hill, New York, 1960.

OBERT, E. F. and YOUNG, R. L., *Elements of Thermodynamics and Heat Transfer*, McGraw-Hill, New York, 2nd ed., 1962.

ROBERTS, J. K., *Heat and Thermodynamics*, Blackie, London, 5th ed., 1960.

SORENSON, H. A., *Principles of Thermodynamics*, Holt, Rinehart & Winston, New York, 1961.

SWALIN, R. A., *Thermodynamics of Solids*, Wiley, New York 1962.

TRIBUS, M., *Thermostatics and Thermodynamics*, Van Nostrand, Princeton, New Jersey, 1961.

WILKS, J., *The Third Law of Thermodynamics*, Oxford University Press, London, 1961.

The American Society of Mechanical Engineers often holds symposia of value on thermodynamics such as the 2nd Symposium on Thermophysical Properties in 1962, and the papers are usually printed, in this case as *Progress in International Research on Thermodynamics and Transport Properties*, American Society of Mechanical Engineers, New York, 1962.

Questions

1. Compare and contrast any two of the general textbooks covering Heat.
2. What different ways are there of measuring temperature?
3. Why is the field of low temperature an interesting one?

CHAPTER 12

Electricity, Magnetism
Dewey 537, 538 Classes

Electricity [Dewey 537 Class]

Electricity involves electric charges and their effect when at rest and when in motion, the major use being in devices using electric currents, and with static electricity in the production of high electric fields, such fields being used in industry as accelerating fields for charged particles and for testing the ability of components to withstand high voltages.

Dictionaries, Books

Aslib have done a valuable service in compiling a handlist of the basic material required to answer enquiries in electrical and electronic engineering, *Handlist of Basic Material for Librarians and Information Officers in Electrical and Electronic Engineering*, Aslib, London, 1960. This includes details of encyclopaedias, dictionaries, handbooks, yearbooks, trade directories, booklists, periodical lists, subject bibliographies, standards, tables, etc.

Very useful for the practical side of electricity is R. L. Oldfield's *A Practical Dictionary of Electricity and Electrons*, Technical Press, London, 1959, for as well as defining fundamental concepts and also more complex terms, with illustrations where necessary, a handbook is included which contains formulae, tables, symbols and circuit diagrams.

Of the many books covering the field one classic treatment is by M. Abraham and R. Becker, *The Classical Theory of Electricity and Magnetism*, Blackie, London, 2nd ed., 1950. This is a translation of Volume 1 of the 14th German edition. The student

124

beginning his studies would be well advised to study H. Cotton's *Electrical Principles*, Cleaver Hume, London, 3rd ed., 1961, which is a beginner's text written entirely in terms of the MKS system of units, and H. E. Hadley's *Magnetism and Electricity for Beginners*, Macmillan, London, 1945. The usual elementary physics course plus elementary electrical engineering courses on circuits and fields are covered by H. G. Booker in *An Approach to Electrical Science*, McGraw-Hill, New York, 1959.

Degree students should be well satisfied with any of the following:

BLEANEY, B. I. and BLEANEY, B., *Electricity and Magnetism*, Oxford University Press, London, 1957.

COULSON, C. A., *Electricity*, Oliver & Boyd, Edinburgh, 1958.

FEWKES, J. H. and YARWOOD, J., *Electricity, Magnetism and Atomic Physics*, University Tutorial Press, London, 1956.

SMITH, C. J., *Electricity and Magnetism*, Arnold, London, 2nd ed., 1959.

STARLING, S. G., *Electricity and Magnetism for Advanced Students*, Longmans, New York, 8th ed., 1953.

Whilst other well-received texts include:

BISHOP, C. C., *Fundamentals of Electricity*, Chilton, Philadelphia, 1960.

BROWN, R. C., *Electricity and Magnetism*, Longmans, London, 1960.

PANOFSKY, W. K. H. and PHILLIPS, M., *Classical Electricity and Magnetism*, Addison-Wesley, Reading, Massachusetts, 2nd ed., 1962.

SCOTT, W. T., *The Physics of Electricity and Magnetism*, Wiley, New York, 1959.

SHIRE, E. S., *Classical Electricity and Magnetism*, Cambridge University Press, London, 1960.

Periodicals and Abstracts

Under the heading electricity, *Ulrich's Periodicals Directory* lists over 100 periodicals dealing with electrical science and tech-

E**

nology, with titles such as *Electrical Review, Journal of the Institution of Electrical Engineers* and *Power Apparatus and Systems*. These periodicals and the other major ones throughout the world are abstracted by such services as *Physics Abstracts* (*Science Abstracts A*), *Electrical Engineering Abstracts* (*Science Abstracts B*), *Physikalische Berichte*, but as well as these, other specialized abstracting bulletins covering electrical science and engineering could be useful. Examples of this type of publication are:

Bibliografia Italiana di Elettrotècnica. Issued bi-monthly by the Centro di Documentazione Elettrotècnica, Padua, Italy. Covers Italian literature for electrical engineering.

Electrical Research Association Abstracts. This is issued weekly by the British Electrical and Allied Industries Research Association, Leatherhead, Surrey. All aspects of electrical engineering and related subjects are covered.

Prehled Technick e Hospodarske Literatury: Energetika a Electrotechnika. Issued monthly by Ustredni Technicka Knihovna CSR, Prague.

Referativnyĭ Zhurnal: Elektrotekhnika. Issued monthly by the Academy of Sciences, Moscow, and covers all aspects of electrical and electronic engineering.

Revue Générale de l'Électricité. Issued monthly by the Comité Électrotechnique Français et Union Technique de l'Électricité, Paris. There is comprehensive coverage and abstracts are of the indicative type. A general index covering Volumes 1–33 has been published.

Technisches Zentralblatt Abteilung Elektrotechnik. Issued monthly by Akademie Verlag, Berlin. Coverage is comprehensive, the abstracts being of the informative type.

Societies and Research Organizations

The two organizations which are very much concerned with electrical science and technology in any country are those covering physics and electrical engineering. In Britain, these are the Institute of Physics and the Physical Society and the Institution of Electrical Engineers. The latter Institution was formed in 1871, though under a different title, with the purpose of promoting the advancement of electrical science and engineering and their applications and to facilitate the exchange of ideas on these topics. The Institution publishes *Physics Abstracts* (*Science Abstracts A*) and *Electrical Engineering Abstracts* (*Science Abstracts B*), the *Journal* and *Proceedings* of the Institution.

The American equivalents of these two organizations, and having the same aims as their counterparts, are the American Institute of Physics and the American Institute of Electrical and Electronic Engineers.

Research organizations concerned with electricity include the American National Bureau of Standards whose electricity division does work on resistance and reactance, electrochemistry, electrical instruments, magnetic measurements, dielectrics, high voltage and absolute electrical measurements, and the British National Physical Laboratory whose recent work includes the calibration of standard equipment, design of an alternating current bridge for the measurement of resistance and the development of thermal converters from which it is possible to obtain an output of about 100 millivolts with only a small temperature rise of the heater element.

The Electrical Research Association at Leatherhead obviously has a large stake in the field of electrical research and works on a variety of topics such as the use of electricity in industry, agriculture and horticulture, studies the insulating, magnetic and structural materials for electrical equipment and the design of such equipment, and the generation, transmission and distribution of electricity. The Association is financed by members' subscriptions and by government grant based on the amount of these subscriptions, so that whilst information is freely available to members it is only available to non-members at the Association's discretion.

As well as the vast amount of research work undertaken at universities other bodies doing research on electrical topics would include the electricity generating organizations, post offices, broadcasting organizations and the armed services.

Electromagnetism [Dewey 537.1 Class]

This is the phenomenon which depends on the interaction of electric and magnetic fields, based on the observations that a charge moving in a magnetic field experiences a force and that a moving electric charge produces a magnetic field. Applications of interest include waveguides, transmission lines and space-charge control tubes.

An early classic work is by O. Heaviside, *Electromagnetic Theory*, Benn, London, 1893–1912, whilst a more recent one which has come to be regarded as a classic is J. A. Stratton's *Electromagnetic Theory*, McGraw-Hill, New York, 1941. Also valuable not least for the good bibliographies contained in it is A. O'Rahilly's *Electromagnetism: a Discussion of Fundamentals*, Longmans, New York, 1938.

R. M. Fano and others develop a consistent macroscopic theory of electromagnetism in their *Electromagnetic Fields, Energy and Forces*, Wiley, New York, 1960. They discuss the relation between circuit theory and field theory and the macroscopic theory is developed in successive steps from the Lorentz force, the integral form of Maxwell's equations in free space and suitable microscopic models of polarized and magnetized matter. Designed to be used with Fano's book is one by H. A. Haus and J. P. Penhune, *Case Studies in Electromagnetism*, Wiley, New York, 1960. This presents problems with solutions and laboratory experiments.

The following is a selection of other useful titles:

ADLER, R. B. *et al.*, *Electromagnetic Energy, Transmission and Radiation*, Wiley, New York, 1960.

CULLWICK, E. G., *Electromagnetism and Relativity with Particular Reference to Moving Media and Electromagnetic Induction*, Longmans, London, 2nd ed., 1959.

LANGMUIR, R. V., *Electromagnetic Fields and Waves*, McGraw-Hill, New York, 1961.

MOULLIN, E. B., *The Principles of Electromagnetism*, Clarendon Press, Oxford, 3rd ed., 1955.

PLONSEY, R. and COLLIN, R. E., *Principles and Applications of Electromagnetic Fields*, McGraw-Hill, New York, 1961.

REITZ, J. R. and MITFORD, F. J., *Foundations of Electromagnetic Theory*, Addison-Wesley, Reading, Massachusetts, 1960.

Electrical and Electronic Measurements

The following are a few of the texts which will give information on methods, equipment and uses:

DUNN, C. H. and BARKER, A. J., *Electrical Measurements Manual*, Prentice-Hall, New York, 1952.

GOLDING, E. W., *Electrical Measurement and Measuring Instruments*, Pitman, London, 4th ed., 1955.

LION, K. S., *Instrumentation in Scientific Research*, McGraw-Hill, New York, 1959.

SMITH, A. W. and WIEDENBECK, M. C., *Electrical Measurement*, McGraw-Hill, New York, 5th ed., 1959.

TERMAN, F. E. and PETTIT, J. M., *Electronic Measurement*, McGraw-Hill, New York, 2nd ed., 1952.

TURNER, R. P., *Basic Electronic Test Instruments, their Operation and Use*, Rinehart, New York, 1953.

Periodicals and Abstracts

All the periodicals covering instrument design and use will at some time have articles of interest on electrical instruments and measurements, periodicals such as *Journal of Scientific Instruments, Review of Scientific Instruments, Instrument Practice, Instrument Engineer, Transactions of the Society of Instrument Technology, Instrument Society of America Journal* and *Instrument Construction*, which is an English translation of the Russian periodical *Priborostroenie*.

These periodicals are abstracted by the specialist publication *Instrument Abstracts*, which is compiled by the British Scientific Instrument Research Association and published monthly by Taylor & Francis, London; they will of course also be in the comprehensive abstracting periodicals. Each issue has an author index and annual subject indexes are published.

Dielectrics [Dewey 537.24 Class]

Dielectrics are materials which are electrical insulators or in which an electric field can be sustained with a minimum of dissipation of power; they are used as insulators and in many devices such as rectifiers and amplifiers. One application of dielectrics—dielectric heating, that is the heating of a nominally electrical insulating material due to its own electrical losses when the material is placed in a varying electrostatic field—is widely used for quick heating of thermosetting glues, foundry-core baking, and jelling and drying of foam rubber.

The National Research Council has published a review of the *Digest of Literature on Dielectrics*, National Research Council, Washington D.C., and using this could be a valuable time saver when searching for information. To keep up to date with current developments periodicals such as *Soviet Physics—Solid State*, which is a cover-to-cover translation of the Russian *Fizika Tverdogo Tela* and contains articles detailing theoretical and experimental investigations in dielectrics, should be perused. A valuable annual review of current knowledge and techniques is *Progress in Dielectrics*, Heywood, London.

Recent books which have been well received include:

BIRKS, J. B. (ed.), *Modern Dielectric Materials*, Heywood, London, 1960.

FRÖHLICH, H., *Theory of Dielectrics, Dielectric Constants and Dielectric Loss*, Oxford University Press, London, 2nd ed., 1958.

KEMP, P., *The Dielectric Circuit*, Chapman & Hall, London, 1960.

Piezoelectricity [Dewey 537.244 Class]

The piezoelectric effect is the phenomenon exhibited by some crystals which when subject to an electric field expand along one axis and contract along another. The converse effect of mechanical strains producing opposite charges on different faces of the crystal also applies. The piezomicrophone and piezoelectric resonator use this effect as their basis of operation.

Reports which were originally prepared for use within the Post Office have been issued, *Piezoelectricity*, HMSO, London, 1957. They deal with the interaction of elastic and electrical properties of crystals with particular reference to the dynamic aspects and to the production and uses of water soluble piezoelectric crystals. The work of the Bell Laboratories leading to the discovery of dipotassiumtartrate and ethylene diamine crystals with favourable properties is described by W. P. Mason in *Piezoelectric Crystals and Their Application to Ultrasonics*, Van Nostrand, New York, 1950. Another useful text is W. G. Cady's *Piezoelectricity*, McGraw-Hill, New York, 1946.

Societies of interest include the American Institute of Radio Engineers Inc. who have a professional group concerned with piezoelectric crystals.

Electronics and Electronic Physics [Dewey 537.5 Class]

This is generally the study and application of electron motion including the theory and means for producing it. Electronic devices are used in communication, television, industrial control and many other applications.

Encyclopaedia and Dictionaries

Useful encyclopaedias include:

International Dictionary of Physics and Electronics, Van Nostrand, Princeton, New Jersey, 2nd ed., 1961. Laws, relationships, equations, basic principles, instruments, components and apparatus are defined. Arrangement is alphabetical and there is a multilingual index in French, German, Russian and Spanish. A useful introduction traces the connection between classical and modern developments.

SARBACHER, R. I., *Encyclopaedic Dictionary of Electronics and Nuclear Engineering*, Pitman, London, 1959. This covers all terms, equipments, elements, components and systems and there is extensive cross-referencing.

SUSSKIND, C. (ed.), *The Encyclopaedia of Electronics*, Reinhold, New York, 1962, which contains over 500 articles arranged alphabetically, with cross-references and a subject index. It covers basic principles, materials, components, systems, industrial applications and manufacturing techniques.

H. Carter has compiled a recent dictionary, *Dictionary of Electronics*, Newnes, London, 1960, which gives concise definitions of terms with appendices giving circuit symbols, abbreviations, colour codes, conversion tables and valve bases.

Books and Bibliographies

The searcher for information on electronic matters is fortunate in that a very excellent bibliography has been compiled by C. K. Moore and K. J. Spencer, *Electronics: a Bibliographical Guide*, Macdonald, London, 1961. The guide covers the period 1945–59 and items are arranged under broad subject headings in date order, both books and periodical articles being quoted. The first section lists bibliographies, buyers' guides, histories, periodicals, abstracts, etc. An author and subject index is also included.

Aslib have considered all the reference material which could be useful for the searcher for electronic information and as a result have published *Handlist of Basic Reference Material for Librarians and Information Officers in Electrical and Electronic Engineering*, Aslib, London, 1960. Included are encyclopaedias, dictionaries, handbooks, yearbooks, trades directories, booklists, periodical lists, subject bibliographies, standards, tables, etc. Somewhat similar is the listing by the British Institution of Radio Engineers which contains a bibliography of relevant books under various subject headings, an article on radio and electronic literature, standards and specifications, and details of international bodies concerned with radio and electronics, *Library Services and Tech-*

nical Information for the Radio and Electronics Engineer, British Institution of Radio Engineers, London, 1959.

Over 10,000 references are contained in W. B. Nottingham's compilation *Bibliography on Physical Electronics*, Addison-Wesley, Cambridge, Massachusetts, 1954. The period covered is mainly 1930–50, but some of the most quoted references after 1950 and in the period 1900–30 are given. References are under the headings gaseous electronics, surface emission and surface phenomena, solid state and conduction, phosphors and luminescence, photocells, photovoltaic effect and photoconductivity techniques.

A review of the scientific and technical information services in Britain of particular interest to the electronics industry, by J. C. Brown, was published in 1955 and may still be valuable, "Information Services on Electronics", *Brit. Commun. Electronics*, **2**, 66 (December 1955).

Recent developments can be followed from the twice-yearly *Advances in Electronics and Electron Physics*, Academic Press, New York, or from the proceedings of conferences such as *Advances in Quantum Electronics*, Columbia University Press, New York, 1961, which covers a conference held at Berkeley, California, in 1961. D. B. Langmuir and W. D. Herschberger edited a review of current thinking and possible future advances which gives a good picture of the state of knowledge in 1961, *Foundations of Future Electronics*, McGraw-Hill, New York, 1961.

A selection from the many recent useful texts follows:

CLARK, D. E. and MEAD, H. J., *Electronic, Radio and Microwave Physics*, Heywood, London, 1961.
HUGHES, R. J. and PIPE, P., *Introduction to Electronics*, English Universities Press, London, 1962.
KLEMPERER, O., *Electron Physics: the Physics of the Free Electron*, Butterworth, London, 1959.
LURCH, E. N., *Fundamentals of Electronics*, Wiley, New York, 1960.
RAMEY, R. L., *Physical Electronics*, Prentice-Hall, London, 1961.
STORM, A. T., *Electronics*, Pitman, London, 2nd ed., 1959.

STEINBERG, W. B. and FORD, W. B., *Electricity and Electronics, Basic*, Technical Press, London, 1957.

Periodicals and Abstracts

There is a wide variety of periodicals covering electronics; indeed *Ulrich's Periodicals Directory* lists over forty and the reader should have access to at least one of these and thus be able to keep up to date with developments in the field. Examples are *Solid State Electronics; Soviet Physics—Solid State*, which is a translation of the Russian *Fizika Tverdogo Tela; Electronics* and *Electronic Technology*. The latter, which was originally *Wireless Engineer* and then *Electronic and Radio Engineer* before assuming its present title, carries abstracts compiled by the Radio Research Station, Slough, grouped under broad subject headings. Annual subject and author indexes are issued. R. England has compiled for Aslib a *Union List of Periodicals on Electronics and Related Subjects*, Aslib, London, 1961.

An attempt was made to establish a comprehensive index to the world's literature and United States patents in the fields of electronics, communications and related engineering as *Electronics Engineering Master Index*. Only four volumes were published covering the years 1925–45, 1946, 1947–48, 1949. Included in each volume is a bibliography of engineering books and declassified United States, British and Canadian documents. Since September 1961, *Electronics and Communications Abstracts* has been published bi-monthly by Multiscience Publishing, Essex. Under broad subject headings the world's major periodical literature, patents, conference proceedings, reports and books are covered.

A list of periodicals with the type of abstract carried was published in 1957 and may still be useful. It was by J. T. Milek, "Abstracting and Indexing Services in Electronics and Related Electrical Fields", *Amer. Docum.*, **8**, 5 (January 1957).

Societies and Research Organizations

British societies interested in electronics include the Institute of Physics and the Physical Society, who have a special subject group on electronics, and the Institution of Electronics. This Institution was initiated in 1930 and has about 1600 members; it holds meetings in various parts of the country, has an annual exhibition in Manchester and publishes quarterly *The Proceedings of the Institution of Electronics*. Also the British Institution of Radio Engineers has as its purpose the advancement of science with particular reference to the development of radio and electronic engineering. Founded in 1925, the Institution holds meetings and publishes a monthly journal.

Comparable American societies include the American Institute of Physics and the Institute of Electrical and Electronic Engineers. The Armed Forces Communications and Electronics Association was organized in 1946 as the Army Signals Association; after various name changes the present name came into being in 1955. It serves to maintain and improve co-operation between the Armed Forces and industry in communications, and operation of communication, electronic and photographic equipment. An annual meeting is held and a monthly journal *Signal* is published.

Research work is undertaken in university and industrial laboratories as well as in governmental laboratories such as the Services Electronics Research Laboratories, Baldock, Radio Research Station, Slough, the National Physical Laboratory, and the Post Office Research Station. European laboratories include the Italian Centro di Studio per l'Elettronica e le Telecomunicazioni.

Electric Discharge and Cathode Ray Tubes
[Dewey 537.52, 537.53 Classes]

Electric discharge is concerned with the passage of electricity through a gaseous, liquid or solid dielectric and this is usually accompanied by luminous effects. Interesting texts covering electric discharge include:

EMELEUS, K. G., *The Conduction of Electricity Through Gases*, Wiley, New York, 3rd ed., 1951.

ENGEL, A. VON and STEENBECK, M., *Elektrische Gasentladungen ihre Physik und Technik*, Springer, Berlin, 1932–34.

LOEB, L. B., *Basic Processes of Gaseous Electronics*, University of California Press, Berkeley, California, 1955. This covers ionic mobilities; diffusion of carriers in gases; velocities of electrons in gases, distribution of energy of electrons in a field of gas (theory and general considerations, theory and use of probes, formation of negative ions; recombination of ions; electrical conduction in gases below ionization by collision; ionization by collision of electrons in a gas) Townsend's first coefficient, the second Townsend coefficient.

MEEK, J. M. and CRAGGS, J. D., *Electrical Breakdown of Gases*, Oxford University Press, London, 1953.

PENNING, F. M., *Electrical Discharge in Gases*, Cleaver-Hume, London, 1957.

Cathode ray tubes are electron tubes where a beam of electrons focused onto a surface and varied in position and intensity on the surface, as in a television tube. Cathode ray oscilloscopes are instruments which are capable of tracing out and giving a visible record of an oscillation.

A recent encyclopaedia by J. F. Rider and S. D. Uslan, *Encyclopaedia on Cathode Ray Oscilloscopes and Their Uses*, Chapman & Hall, London, 2nd ed., 1959, should prove most useful. It has information on what an oscilloscope is and what it can do and how to use it for particular applications. A subject index and bibliography are included. Textbooks of value include:

CARTER, H., *An Introduction to the Cathode Ray Oscilloscope*, Cleaver-Hume, London, 1957.

PARR, G. and DAVIE, O. H., *Cathode Ray Tube and Its Applications*, Cleaver-Hume, London, 3rd ed., 1959.

REYNER, J. H., *Cathode Ray Oscillographs*, Pitman, London, 5th ed., 1957.

Ionization [Dewey 537.532 Class]

Ionization is the process which results in the formation of ions. In gases or dielectric liquids, this follows the passage of particles of sufficient energy, absorption of radiation, attainment of sufficiently high temperature, or application of a strong electric field. Ionization chambers are used to detect radiations and particles, whilst ionization gauges measure vacuum.

Conference proceedings are always useful means of keeping up to date and recent ones which are available for consultation include:

MAECKER, H. (ed.), *Proceedings of the Fifth International Conference of Ionization Phenomena in Gases, Munich, 28th August–1st September, 1961*, North-Holland, Amsterdam, 1962.

G. Francis has written an interesting book which is intended to be complementary to other books in the field in that it treats those branches of the subject which are already closely linked with other branches of physics, such as atmospheric phenomena linked with astrophysics, high frequency discharges and radio wave propagation, and high current discharges with nuclear physics and electrodynamics. It is *Ionization Phenomena in Gases*, Butterworth, London, 1960. In the main, it is intended for those finishing their degrees or about to begin research. Other useful books include:

CROWTHER, J. A., *Ions, Electrons and Ionizing Radiations*, Arnold, London, 8th ed., 1949.

DELCROIX, J. L., *Introduction to the Theory of Ionized Gases*, Interscience, New York, 1960.

MASSEY, H. S. W. and BURHOP, E. H. S., *Electronic and Ionic Impact Phenomena*, Oxford University Press, London, 1952.

X-Rays

[Dewey 537.535 Class (See also X-Ray Diffraction 548.83)]

X-Rays are electromagnetic radiations of about 1Å wavelength which are emitted from a discharge tube in which cathode rays are

138   HOW TO FIND OUT ABOUT PHYSICS

allowed to fall on solid matter. The uses of X-rays include radiography, tumour destruction and the examination of matter on an atomic scale.

An interesting simple "history" of X-rays by A. R. Bleich is *The Story of X-Rays from Röntgen to Isotopes*, Dover, New York, 1960, whilst valuable texts include:

COMPTON, A. H. and ALLISON, S. K., *X-Rays in Theory and Experiment*, Macmillan, London, 2nd ed., 1935.

SCHALL, W. E., *X-Rays: their Origin, Dosage and Practical Application*, Wright, Bristol, 8th ed., 1961.

SELMAN, J., *The Fundamentals of X-Ray and Radium Physics*, Thomas, Springfield, Illinois, 2nd ed., 1957.

Photoconductivity, Photoelectricity, Thermionics and Secondary Electron Emission
[Dewey 537.54, 537.533 Classes]

Photoconductivity is that property of certain materials which vary their electrical conductivity under the influence of light, whilst photoelectricity is electricity produced by the action of light and whose applications include the photoelectric cell and scintillation counter. Thermionics strictly deals with the emission of electrons from hot bodies, but is generally used in the broader sense of the behaviour and control of such electrons; applications include high-voltage cathode rectifiers. Secondary electron emission is electron emission from a surface which is bombarded by electrons from another source.

Texts of value to students of these subjects include:

BRUINING, H., *Physics and Applications of Secondary Electron Emission*, Pergamon, London, 1954.

BUBE, R. H., *Photoconductivity of Solids*, Wiley, New York, 1960.

GRAY, D. E., *The Dynamic Behaviour of Thermoelectric Devices*, Wiley, New York, 1960.

HEIKES, R. R. and URE, R. W. (eds.), *Thermoelectricity: Science and Engineering*, Interscience, New York, 1961.

HUGHES, A. L. and DuBRIDGE, L. A., *Photoelectric Phenomena*, McGraw-Hill, New York, 1932. This is an exhaustive review of photoconductivity up to 1930.

KAYE, J. and WELSHE, J. A. (eds.), *Direct Conversion of Heat to Electricity*, Wiley, New York, 1960.

MACDONALD, D. K. C., *Thermoelectricity: an Introduction to the Principles*, Wiley, New York, 1962.

McGEE, J. D. and WILCOCK, W. L. (eds.), *Photo-Electric Image Devices: Proceedings of a Symposium held in London, September, 1958*, Academic Press, New York, 1960.

Electron Optics [Dewey 537.56 Class]

Electron optics is concerned with the control of the motion of electrons. The name arises because the mathematical formulation is the same as for light in refracting media. Cathode ray tubes and electron microscopes are the main applications. The electron microscope, a most valuable laboratory instrument, consists of a thermionic tube where the electrons emitted from the cathode are focused, using an electrostatic field, to form an enlarged image of the cathode on a fluorescent screen.

The literature has been surveyed by V. E. Coslett and C. Marton and others in two bibliographies which will prove good starting points for anyone studying the subject. Coslett's bibliography, *Bibliography of Electron Microscopy*, Longmans, London, 1951, covers the period 1927–48 and Marton's *Bibliography of Electron Microscopy*, U.S. Government Printing Office, Washington D.C., 1950, the period 1926–49. For later information the New York Society of Electron Microscopists publication *Bibliography of Electron Microscopy* can be consulted.

Amongst recent useful texts are the following:

HAINE, M. E., *The Electron Microscope: the Presentation of the Art*, Spon, London, 1961.

KAY, D. (ed.), *Techniques for Electron Miscroscopy*, Pergamon, London, 1961.

WISCHNITZER, S., *Introduction to Electron Microscopy*, Pergamon, London, 1962.

WYCKOFF, R. W. G., *The World of the Electron Microscope*, Yale University Press, New Haven, Connecticut, 1958.

Abstracts

The major abstracting services cover electron microscopy and it was covered by *Electron Microscopy Abstracts*, which was reprinted from the *Journal of the Royal Microscopical Society*, and issued as a separate series approximately four times a year. Publication commenced in 1949 and ceased in March 1961.

Societies and Research Organizations

American societies with the special purpose of increasing and diffusing the knowledge of electron microscopy, its applications and results of research, are the Electron Microscope Society of America which was first organized in 1942, and the New York Society of Electron Microscopists which was first organized in 1951. In Britain the Royal Microscopical Society and the Institute of Physics and the Physical Society are concerned with electron microscopy, and in Germany the relevant society is the Deutsche Gesellschaft für Elektronenmikroskopie.

Research work is done by the manufacturers of electron microscopes amongst others, whilst research work with the microscope is done by universities, technical colleges and many other industrial and governmental laboratories such as the National Physical Laboratory, Italian Instituto Superiore della Poste et delle Telecomunicazioni, Norwegian Sentralinstitutt för Industriell Forskning, and the German Institut für Elektronenmikroskopie.

Electrodynamics [Dewey 537.6 Class]

Electrodynamics is concerned with the forces exerted on a conductor carrying a current by a current flowing in another conductor. The action is entirely magnetic and was first noticed by Ampère.

Recent texts which may be consulted include:

JACKSON, J. D., *Classical Electrodynamics*, Wiley, New York, 1962.

LANDAU, L. D. and LIFSHITZ, E. M., *Electrodynamics of Continuous Media*, Pergamon, London, 1960.

MARN, P. and SPENCER, D. E., *Foundations of Electrodynamics*, Van Nostrand, Princeton, New Jersey, 1960.

Masers [Dewey 537.6 Class]

Maser is short for Microwave Amplification and Oscillation by Stimulated Emission of Radiation. Masers have also operated at ultra-high frequencies and can in principle operate at infrared and optical frequencies. At present, it is valuable in fields such as radar and radioastronomy and one of its valuable characteristics is that a maser amplifier has very low noise.

J. R. Singer has written a good text from both the classical and quantum mechanical points of view, *Masers*, Wiley, New York, 1959. Included is a description of the ammonia maser, a theoretical discussion of a magnetic atomic beam system, optically pumped frequency standard, electron paramagnetic resonance, two-level masers, three-level cavity masers, theory and experimental results of the travelling wave, maser design techniques.

Other recent books include:

TROUP, G., *Masers*, Methuen, London, 1959.

VUYLSTREKE, A. A., *Elements of Maser Theory*, Van Nostrand, Princeton, New Jersey, 1960.

Lasers, which are optical masers and give a beam of highly monochromatic light, are covered by an excellent text written by B. A. Lengyel, *Lasers—Generation of Light by Stimulated Emission*, Wiley, New York, 1962.

Most of the information on lasers appears in technical periodi-

cals and the references can be found from a very useful bibliography prepared by K. J. Spencer, *Lasers*, Ministry of Aviation,
Technical Information and Library Services, London, which
covers the period December 1958–December 1962.

The present uses of lasers are in radar and space communications, but some other applications in medical, chemical and biological fields should follow.

Many of the foremost university and industrial laboratories
throughout the world are working in this field, laboratories such
as the American Bell Telephone Laboratories and the Columbia
Radiation Laboratory and in Britain International Research and
Development Ltd. It has been estimated that eighteen months after
the discovery of the laser 400 laboratories were working in the
field.

Semiconductors [Dewey 537.622 Class]

Semiconductors are materials whose electrical resistance is of
the order of 10^{-2} to 10^9 ohm/cm and thus fall in between those
which readily conduct electricity and those which do not. They
are crystalline materials and exhibit a negative temperature
co-efficient of resistance over most of the temperature range.
Applications of semiconductors are mainly in rectification and
amplification (transistors) and research is undertaken in industry
and by governmental and university laboratories such as the
Russian Institute of Semiconductors.

Current thinking and recent developments in semiconductor
science can be followed by using review publications such as
Progress in Semiconductors, Heywood, London. Volume 6 which
was published in 1962 covered plastic deformation in semiconductors, physics of semiconductor switching devices, magnetic
resonance in semiconductors and excitation spectra in semiconductors and ionic compounds. Proceedings of conferences such as
the International Conference on Semiconductors are usually valuable. The 3rd Conference was held at the University of Rochester
in 1958 and covered band theory, recombination and impurity

centres, surfaces, dislocations, excitons and photons, magneto-optical effects, semiconducting compounds and large-gap semi-conductors. The proceedings were published as *Advances in Semiconductor Science*, Pergamon, New York, 1959. Other con-ferences whose proceedings have been published include the International Conference on Semiconductor Physics held in Prague in 1960, and a conference held in Boston in 1959 on the properties of elemental and compound semiconductors.

Bibliographies, Books and Tables

Several bibliographies have been published including one com-piled by N. L. Meyrick, *Fifteen Years of Semiconducting Materials and Transistors*, Newmarket Transistor, Newmarket, 1958. The bibliography has four parts, semiconductors—theory and measure-ment, processing of semiconductors, rectifiers and diodes, tran-sistors, under which 2511 items are grouped. Earlier bibliographies include one published by the Science Museum in London in September 1952, *Semiconductors* (Science Library Bibliographical Series No. 711), which contains 431 items covering the period 1946–52, and one published by the Admiralty Centre for Scientific Information and Liaison, London, *A Bibliography of Literature Relating to Semiconductors* (ACSIL Library Bibliography No. 8). Published in 1954 this was issued in three parts covering 1948–52, prior to 1948 and current items to August 1952. About 1200 items are included.

As a guide to the various publications of interest to those searching for information on semiconductors E. H. C. Driver has compiled a survey of guides to the literature, abstracts, reviews and surveys, definitions, standards and also includes a selected book list in his *Semiconductors and Transistors*, Library Associa-tion, London, 1962.

Tables available include *Selected Constants Relative to Semi-conductors*, Pergamon, London, 1961. Included are semiconduct-ing properties, physiochemical properties and in some cases extrinsic properties.

As a general survey of the subject, and containing ample

references, W. C. Dunlap's *An Introduction to Semiconductors*, Wiley, New York, 1957, can be recommended. There are very many other useful texts and the following are some of the recent ones:

BLACKEMORE J., *Semiconductor Statistics*, Pergamon, London, 1962.

DRABBLE, J. R. and GOLDSMID, H. J., *Thermal Conduction in Semiconductors*, Pergamon, London, 1961.

HANNAY, N. B. (ed.), *Semiconductor*, Reinhold, New York, 1959.

HILSUM, C. and ROSE-INNES, A. C., *Semiconducting III–V Compounds*, Pergamon, London, 1961.

HUNTER, L. P. (ed.), *Handbook of Semiconductor Electronics*, McGraw-Hill, New York, 2nd ed., 1962.

IOFFE, A. F., *Physics of Semiconductors*, Infosearch, London, 1960. This is a translation of the 1957 Russian edition.

SHIVE, J. N., *The Properties, Physics and Design of Semiconductor Devices*, Van Nostrand, Princeton, New Jersey, 1959.

SMITH, R. A., *Semiconductors*, Cambridge University Press, London, 1959.

TURNER, R. P., *Semiconductors Device*, Rinehart, New York, 1962.

Periodicals and Abstracts

Periodicals which cover semiconductors include the bi-monthly *Solid-State Electronics* which is an international journal serving the semiconductor industry and whose issues include papers reporting original work in various areas of applied solid state physics including transistor technology and also papers on the application of semiconductors, design and performance of galvomagnetic devices. Reporting Russian work is *Fizika Tverdogo Tela*, which is published as a cover-to-cover translation *Soviet Physics—Solid State* by the American Institute of Physics. In this are reported experiments and theoretical investigations on semiconductors and dielectrics. Other periodicals can be found from the various published guides.

Specialized abstracting services include *Solid State Abstracts* which has been issued monthly since 1957 by the Cambridge Communications Corporation, Cambridge, Massachusetts. Formerly known as *Semiconductor Electronics*, the coverage includes U.S. patents, conference reports and theses and there are author and subject indexes in each issue. Battelle Memorial Institute, under the auspices of the Electrochemical Society Inc., compile *Semiconductor Abstracts* and these are published by Wiley, New York. They are issued annually with each issue carrying an author and subject index. Volume 7 covering 1959 was the latest to be published.

Superconductivity [Dewey 537.623 Class]

Superconductivity is the property of some materials where below a certain temperature, usually a few degrees above absolute zero, the electrical resistance of the material becomes immeasurably small. This effect has been utilised in the superconducting bolometer and in devices for switching in computers.

The field is reviewed by C. T. Lane in his *Superfluid Physics*, McGraw-Hill, New York, 1962, the various steps in the development of the subject being outlined and comment included on its present stage of development. An earlier book which contains a very complete set of references is D. Shoenberg's *Superconductivity*, Oxford University Press, London, 2nd ed., 1952. Also of value are:

BREMER, J. W., *Superconductive Devices*, McGraw-Hill, New York, 1962.

LAUE, M. VON, *Theory of Superconductivity*, Academic Press, New York, 1952.

LYNTON, E. A., *Superconductivity*, Methuen, London, 1962.

Magnetism [Dewey 538 Class]

Magnetism is concerned with the laws and conditions of magnetic force and its effects and is the property of certain bodies to attract or repel other bodies. Substances can be grouped as to their behaviour in magnetic fields as paramagnetic, diamagnetic, ferromagnetic and anti-ferromagnetic.

In the bibliography compiled by C. K. Moore and K. J. Spencer, *Electronics: a Bibliographical Guide*, Macdonald, London, 1961, there are two relevant sections, section 12 which covers magnetism and ferromagnetism and has 42 references to books and periodical articles, and section 13 which covers magnetic materials and has 30 references to books and periodical articles.

Conferences are held periodically on magnetism and magnetic materials, the sixth one with that title being held in New York in 1960. The conference covered ordered spin systems, computer devices, spin configurations, metallic films, nuclear hyperfine fields, ferromagnetic resonance, high coercive materials, exchange interactions and non-linear microwave processes, magnetization processes and fine particles, anisotropy, domain walls and domain wall motion, microwave devices, metals and alloys, soft magnetic materials and oxides.

As an introductory text for newcomers to the subject a book by H. E. Hadley can be recommended, *Magnetism and Electricity for Beginners*, Macmillan, London, 1945, whilst for degree students suitable works include:

BARR, A. E. DE, *The Magnetic Circuit*, Institute of Physics, London, 1953.

BARR, A. E. DE, *Soft Magnetic Materials Used in Industry*, Institute of Physics, London, 1953.

BLEANEY, B. I. and BLEANEY, B., *Electricity and Magnetism*, Oxford University Press, London, 1957.

STARLING, D., *Electricity and Magnetism for Advanced Students*, Longmans, New York, 8th ed., 1953.

The various magnetic materials, such as ferrites, garnets, iron and silicon iron alloys, nickel–iron and other alloys, and their properties are discussed in two recent books, F. Brailsford's *Magnetic Materials*, Methuen, London, 3rd ed., 1960, and K. J. Standley's *Oxide Magnetic Materials*, Clarendon Press, Oxford,

1962, whilst permanent magnets and their applications are covered by a book edited by D. Hadfield, *Permanent Magnets and Magnetism*, Iliffe, London, 1962, and R. J. Parker and R. J. Studders, *Permanent Magnets and Their Applications*, Wiley, New York, 1962.

The texts quoted above are some of the well known, or more specific ones, and certainly others such as any of the following are equally valuable:

BATES, L. F., *Modern Magnetism*, Cambridge University Press, London, 4th ed., 1961.

BROWN, R. C., *Electricity and Magnetism*, Longmans, London, 1960.

PANOFSKY, W. K. H. and PHILLIPS, M., *Classical Electricity and Magnetism*, Addison-Wesley, Reading, Massachusetts, 2nd ed., 1962.

SMITH, C. J., *Electricity and Magnetism*, Arnold, London, 2nd ed., 1959.

Research Organizations

Research is undertaken by such as the Permanent Magnet Association Central Research Organization, Sheffield, whose research activities centre mainly on the developing of new permanent magnet materials and economizing in processing of existing alloys, though work on high coercivity materials and new methods of analysis for the complex alloys now used are also activities of the Organization.

Questions

1. Compare the coverage of *Electrical Engineering Abstracts* and *Physics Abstracts* for any aspect of electricity.
2. Compare the various abstracting sources covering semiconductor physics.
3. What are the uses of masers?

F

CHAPTER 13

Atomic, Molecular and Nuclear Physics
Dewey 539 Class

Crystallography and X-Ray Diffraction
Dewey 548, 548.83 Classes

ATOMIC physics covers the study of the physical properties of atoms when the atom is regarded as a whole, i.e. as a nucleus with which is associated a number of electrons. The field is wide and some of the topics which can be classed as atomic physics include atomic spectra, radiations from atoms, interaction of particles and rays with atoms, and the bulk properties of metals.

For the layman seeking to know something of atomic physics there are several books of value. *Explaining the Atom*, Gollancz, London, 1955, was written by S. Hecht and explains the experimental and theoretical advances leading from the earliest concepts of the nature of substances to the release of atomic energy. A valuable introduction is J. M. Valentine's *Teach Yourself Atomic Physics*, English Universities Press, London, 1960. The reader is taken through a series of simple experiments, which duplicate the discoveries of the last fifty years, by O. R. Frich in his book *Meet the Atoms*, Wyn, New York, 1947. G. L. Bush and A. A. Silvidid simplify the nature of the atom in their *The Atom: a Simplified Description*, Barnes & Noble, New York, 1961.

The serious student has many books from which to chose, including classic texts which have proved their worth, such as that by Sir George Thomson, *The Atom*, Oxford University Press, London, 6th ed., 1962. This edition has been brought up to date by the inclusion of material released at the 1954 Geneva Conference on the Peaceful Use of Atomic Energy. Other classics include:

ANDRADE, E. N. DA C., *Structure of the Atom*, Bell, London, 4th ed., 1934.

BORN, M., *Atomic Physics*, Blackie, London, 7th ed., 1962.

With so many valuable books on the subject selection of a few is a well-nigh impossible task, so the following list should be taken to be merely some of the useful titles:

BLACKWOOD, O. H. *et al.*, *An Outline of Atomic Physics*, Wiley, New York, 3rd ed., 1955.

COPELAND, P. L. and BENNETT, W. F., *Elements of Modern Physics*, Oxford University Press, New York, 1961.

JORGENSON, E. K., *Orbitals in Atoms and Molecules*, Academic Press, New York, 1962.

OLDENBERG, O., *Introduction to Atomic and Nuclear Physics*, McGraw-Hill, New York, 3rd ed., 1961.

SLATER, J. C., *Quantum Theory of Atomic Structures*, McGraw-Hill, New York, 1960.

TOLANSKY, S., *Introduction to Atomic Physics*, Longmans, London, 4th ed., 1956.

WEHR, M. R. and RICHARDS, J. A., *Introductory Atomic Physics*, Addison Wesley, Reading, Massachusetts, 1962.

As would be expected from such a wide field many periodicals have articles of interest and details can be obtained from any of the guides mentioned in previous chapters. Those which pay particular emphasis on molecular and atomic physics include the bi-monthly *Molecular Physics* and *Acta Physica Scientiarum*, which is issued irregularly.

Research work on atomic physics is being undertaken by almost every university and by very many industrial and other research organizations, such as the Nordic Institute for Theoretical Atomic Physics, Copenhagen. Most of these bodies will also be doing work on nuclear physics and so the details given in the next section on nuclear physics are relevant. So-called atomic energy, really correctly nuclear energy, will also be dealt with under nuclear physics.

Nuclear Physics [Dewey 539.7 Class]

Nuclear physics is the study of the properties of atomic nuclei, and even though intensive research work has been done, especially during the last twenty years, the nature of the nuclear forces are still not completely understood.

Tables, Handbooks, Dictionaries

Various compilations of experimental and theoretical information exist in the field of basic nuclear physics, and, if consulted, can save the seeker of information much valuable time. R. C. Gibbs and K. Way bring out the existence and character of the various compilations, though no attempt is made to characterize any of the tables as containing the current "best" values, in *A Directory to Nuclear Data Tabulations*, National Academy of Sciences, Washington D.C., 1958.

Most valuable are *Nuclear Data Tables*, National Academy of Sciences, Washington D.C. Publication commenced in 1959, the 1959 tables containing collections of data on charged particle scattering, experimental ground state Q values, isotopic abundances and nuclear moments. The 1960 tables computed Q values, gamma-ray energies, beta disintegration energies, nuclear radii, experimental Q values and nuclear moments.

O. R. Frisch has edited *The Nuclear Handbook*, Newnes, London, 1958, and it is a publication which has earned much praise. Details are given in the form of write-ups, graphs and charts of some concepts in nuclear physics, radiation effects and protection, elements and isotopes, natural radioactivity, materials, vacuum, particle accelerators, charged particles, X-rays and gamma-rays, neutrons, fission product and transuranic elements, reactors, chemistry, ion chambers and counters, electronics, deflection techniques and magnetic materials, cloud chambers and bubble chambers, nuclear emulsions and nuclear reactions. Giving similar information as part of its coverage is a standard work by S. Glasstone, *Sourcebook on Atomic Energy*, Van Nostrand, Princeton, New Jersey, 2nd ed., 1958.

Illustrative of the other kinds of tables available are:

DZHELEPOV, B. S. and PEKER, L. K., *Decay Schemes of Radioactive Nuclei*, Pergamon, London, 1961.

NIJGH, G. J. *et al.*, *Nuclear Spectroscopy Tables*, North-Holland, Amsterdam, 1959.

The very welcome work by W. Kunz and J. Schintlmeister, *Nuclear Tables Part 1: Nuclear Properties of the Elements*, Pergamon, London, 1963, is to be issued in two volumes including all the data concerning stable and radioactive nuclei as well as all decay schemes. Part 2 when issued will contain data on cross-sections, excitation functions, threshold values, Q values, measurements of angular distribution and nuclear levels. Most of the data has been compiled by reference to the original source and compared with other tables.

Dictionaries and glossaries are fairly profuse and include one compiled by the British Standards Institute, *Glossary of Terms Used in Nuclear Science*, British Standards Institute, London, 1962 (British Standard 3455), and one by the United Kingdom Atomic Energy Authority, *A Glossary of Atomic Terms*, HMSO, London, 4th ed., 1963. Another recent one covering English, French, German and Russian terms is by R. Sube, *Dictionary of Nuclear Physics and Technology*, Pergamon, London, 1962. At least one of these dictionaries, or one of the following, should be available in any good library:

A Glossary of Terms in Nuclear Science and Technology, American Society of Mechanical Engineers, New York, 1955.

CHARLES, V., *Dictionnaire Atomique*, Librairie Hachette, Paris, 1960.

HOCKER, W. H. and WEIMER, K., *Lexikon der Kern- und Reaktortechnik*, Franckh'sche Verlagshandlung, Stuttgart, 1959.

SARBACHER, R. I., *Encyclopaedic Dictionary of Electronics and Nuclear Engineering*, Pitman, London, 1959.

Books

Keeping up to date with progress in nuclear physics is quite a task because of the amount of work being undertaken. The various annual reviews, which consider recent research and the development of new ideas, help enormously by bringing all the facts together, editing them and presenting them in a coherent form. Amongst the reviews which are published are:

Advances in Nuclear Science and Technology, Academic Press, New York.

Annual Reviews of Nuclear Science, Annual Reviews, Stanford, California.

Progress in Nuclear Physics, Pergamon, London.

Soviet Reviews of Nuclear Science, Pergamon, London.

Conferences embracing nuclear physics are many. Useful information on current practice and ideas and possible future developments can usually be obtained from the proceedings of these conferences. To mark Rutherford's jubilee a conference was held in Manchester in 1961 which covered the fields of high-energy investigations of nuclei, collective motion in nuclei, the nuclear ground state, direct interactions and weak interactions. Interesting as this conference must have been to the participants, its value is greatly increased by the publishing of the proceedings as *Proceedings of the Rutherford Jubilee International Conference, Manchester, 1961*, Heywood, London, 1961.

Details of the publications covering the many other conferences will be found in library catalogues, abstracting publications and the like. Sufficient here to indicate one or two conference publications of interest:

Proceedings of the 1954 Glasgow Conference on Nuclear and Meson Physics, Pergamon, London, 1955.

Proceedings of the International Conference on Nuclear Structure, Kingston, Canada, 29th August–3 September, 1960, North-Holland, Amsterdam, 1960.

Proceedings of the International Conference on Nuclidic Masses, McMasters University, Hamilton, Canada, 12–16 September, 1960, University of Toronto, Toronto, 1960.

Proceedings of the 1960 Annual International Conference on High Energy Physics, at Rochester, New York, 25 August–1 September, 1960, Interscience, New York, 1960.

Nuclear Forces and the Few Nucleon Problem. Proceedings of a Conference held in London, 8–11 July, 1959, Pergamon, London, 1959.

The two United Nations Conferences on the Peaceful Uses of Atomic Energy, the proceedings of which have both been published, contained much information of interest on nuclear physics and should not be overlooked.

Again general textbooks are many and those based on actual courses are of proved value. Such is R. D. Evans's *The Atomic Nucleus,* McGraw-Hill, New York, 1955, which is based on a course given at the Massachusetts Institute of Technology and whose approach is between that of a theoretical treatise and a detailed handbook of experimental techniques. Similarly, M. G. Mayer and J. Orear have produced a book from lectures, *Nuclear Physics: a Course given by Enrico Fermi at the University of Chicago,* Chicago University Press, Chicago, 1950.

The following list of books should be taken as indicative rather than selective:

ASIMOV, I., *Inside the Atom,* Abelard-Schuman, London, 1956.

BETHE, H. A. and MORRISON, P., *Elementary Nuclear Theory,* Wiley, New York, 2nd ed., 1956.

EISENBUD, L. and WIGNER, E. P., *Nuclear Structure,* Oxford University Press, London, 1958.

ELTON, L. R. B., *Nuclear Sizes,* Oxford University Press, London, 1961.

FEENBERG, E., *Shell Theory of the Nucleus,* Princeton University Press, Princeton, New Jersey, 1955.

HALLIDAY, D., *Introductory Nuclear Physics,* Wiley, New York, 2nd ed., 1955.

HEISENBERG, W., *Nuclear Physics,* Methuen, London, 1953.

OLDENBERG, O., *Introduction to Atomic and Nuclear Physics,* McGraw-Hill, New York, 3rd ed., 1961.

PRESTON, M. A., *Physics of the Nucleus*, Addison-Wesley, Reading, Massachusetts, 1962.

Periodicals

Periodicals of interest include the fortnightly *Nuclear Physics*, which covers the experimental and theoretical study of the fundamental constituents of matter and their interactions, the monthly *Nucleonics*, and the German monthly *Nukleonik*.

Abstracting publications covering nuclear physics include *Nuclear Science Abstracts* and *Physics Abstracts* (*Science Abstracts A*).

Nuclear Science Abstracts is issued bi-monthly by the United States Atomic Energy Commission, each issue carrying an author and subject index. Frequent cumulations are issued. It includes technical reports of the USAEC and its contractors, technical reports of U.S. government agencies and industrial and research organizations as well as patents, books and world-wide journal literature.

Societies and Research Organizations

The major physics society in a particular country will have an interest in nuclear physics and may have set up a section for its interested members as the New York Academy of Sciences has done. The Academy stimulates the advancement of knowledge by holding meetings and publishing its transactions and also by offering the Boris Pregel Award for the most acceptable paper embodying the results of research in nuclear physics and nuclear engineering.

All the major countries throughout the world have governmental research organizations concerned with nuclear power and research and thus will work in the various fields of nuclear physics. The majority of the work is now free from any security restrictions and so these organizations, and the various nuclear engineering societies, should prove useful sources of information. Details will be given in the next section on nuclear energy.

The National Institute for Research in Nuclear Science in

Britain is working in the main field of elementary particle physics, and details of this and other work of the laboratory can be found in its annual report.

Universities at the present time are particularly strong on nuclear research work and details of their current work can be found from the guides mentioned in chapter seven.

Nuclear Energy, Atomic Energy

Nuclear energy is the release of energy when fission occurs in the nucleus of uranium or plutonium. A considerable amount of research and development work has been undertaken in the last twenty years to harness this energy so that it becomes an economical electricity generating unit.

Useful short guides to information sources on atomic energy are:

ANTHONY, J. L., *Sources of Information on Atomic Energy*, (AERE-Lib/L1 3rd ed.), HMSO, London, 1960. This discusses sources of information in various countries and gives a bibliographical listing of published sources of information.

WOOD, A. J., *A Guide to Information on Atomic Energy*, HMSO, London, 1960. This lists general bibliographies, books and periodicals and specialized publications dealing with the power programme and nuclear engineering, radioisotopes, nuclear physics, radiation hazards and protection and legislation.

The United Nations published an early bibliography on atomic energy, *An International Bibliography on Atomic Energy*, United Nations, New York, 1951, plus supplements. Volume 1 covers political, social and economic aspects, and Volume 2 scientific aspects. Volume 2 and its supplements is divided into five sections: fundamental nuclear science, physics and engineering of nuclear energy, biological and medical effects of high energy radiation, isotopes in biology and medicine, applications of nuclear physics in non-biological sciences and technologies. Other bibliographies on various aspects of atomic energy are issued by the United States Atomic Energy Commission, United Kingdom Atomic Energy Authority, International Atomic Energy Agency and other bodies. They will all be included in *Nuclear Science Abstracts*.

For a simple introduction T. E. Allibone's *The Release and Use of Atomic Energy*, Chapman & Hall, London, 1961, can be recommended. It is based on a Royal Institution Christmas lecture. Other good introductions are:

ATKINSON, W. G., *Introduction to Atomic Energy*, Rider, New York, 1959.

CHAPMAN, R., *Atomic Energy for All*, Odhams, London, 1960.

Of the many excellent books covering the whole of atomic energy those edited by S. Glasstone can be recommended, such as *Principles of Nuclear Reactor Engineering*, Macmillan, London, 1956, and *Sourcebook on Atomic Energy*, Van Nostrand, Princeton, New Jersey, 2nd ed., 1958. Others of the same standard are the *Nuclear Engineering Handbook*, McGraw-Hill, New York, 1958, which is edited by H. Etherington and contains information on reactor physics, control, isotopes, materials, radiation protection, etc., and the *Reactor Handbook*, McGraw-Hill, New York, 2nd ed., 1960. This is issued in several volumes including one on physics which has information on reactor physics, nuclear physics, reactor statics, neutron attenuation, etc.

Many other books are available covering both general and specific aspects and many are published as proceedings of conferences, such as the conference held in Vienna in 1961 under the auspices of the International Atomic Energy Agency whose proceedings were published as *Physics of Fast and Intermediate Reactors*, IAEA, Vienna, 1962. Reviews of all aspects of atomic energy are contained in the proceedings of the two *United Nations Conferences on the Peaceful Uses of Atomic Energy*. *Progress in Nuclear Energy*, Pergamon, London, is published in twelve separate parts and reviews all aspects of atomic energy. Series 1 deals with physics and mathematics.

Periodicals

The major periodicals in the field are *Nucleonics*, *Nuclear*

Engineering, Journal of Nuclear Energy, the German *Kerntechnik* and the Russian *Atomnaya Energiya.* All of these are abstracted by *Nuclear Science Abstracts,* which is the major abstracting periodical for atomic energy. *Nuclear Engineering Abstracts* is published by Silver End Documentary Publications, London, and contains indicative abstracts of articles in periodicals and newspapers.

Societies and Research Organizations

The American Nuclear Society was formed in 1954 with the object of integrating and advancing nuclear science and engineering and it does this by holding meetings and publishing the *Transactions* of the Society and *Nuclear Science and Engineering.* Another American organization in this field is the Atomic Industrial Forum, and its aims are to encourage the development and utilization of atomic energy by holding conferences on all aspects of this theme. The British Nuclear Energy Society provides a forum for discussion on all aspects of nuclear energy by holding meetings and publishing original papers in its *Journal.* All the major societies whose work has any bearing on nuclear energy are members of the Society.

The national atomic energy authority of each country details some of its research work in its annual report. The United Kingdom Atomic Energy Authority reports on the work being pursued in nuclear physics, theoretical physics, physics of the solid state and radiation damage, in its annual report. The United States Atomic Energy Commission summarizes some of the more scientifically interesting advances being made under its fundamental research programme in its annual research report. The fourth annual report appeared as *Fundamental Nuclear Energy Research, 1963,* U.S. Government Printing Office, Washington D.C., 1963.

International bodies include the International Atomic Energy Agency, Vienna; EURATOM, Brussels; and OECD European Nuclear Energy Agency, Paris, all of whom do research work and dispense information. As well as these there are many industrial laboratories of firms, who build nuclear power plants or produce equipment, that are doing research in this field. Details can be found from the *World Nuclear Directory*, Harrap, London, 2nd ed., 1963.

Neutron Physics, Neutron Optics, Elementary Particles, Cosmic Rays, Mesotrons, etc.
[Dewey 539.721, 539.722/3 Classes]

Neutron physics covers all aspects of neutron behaviour whilst neutron optics is that aspect where the wave characteristics dominate and lead to behaviour similar to that of light.

Neutron physics has been the subject of a recent symposium at the Reneselaer Polytechnic Institute whose proceedings have been edited by M. C. Yeater and appear as *Neutron Physics*, Academic Press, New York, 1962. The principal subjects covered are the physics of nuclear reactors and the neutron physics fundamental to that field, but the work will also interest those concerned with the physics of crystalline solids and nuclear energy levels at or above the binding energy. Contributions include pulsed neutron studies, neutron cross-sections, neutron transport, neutron spectrometry and plasma research.

L. F. Curtiss has written a standard *Introduction to Neutron Physics*, Van Nostrand, Princeton, New Jersey, 1959, which includes an introduction and information on particle and nuclear interactions, sources, detection of neutrons, spectrometers and monochromators, interactions of neutrons with matter, calibrations and standards, neutron shielding and protection of personnel, and formulas of neutron physics.

Other useful books include:

BACON, G. E., *Neutron Diffraction*, Oxford University Press, London, 2nd ed., 1962.

DAVISON, B., *Neutron Transport Theory*, Clarendon Press, Oxford, 1957.

HUGHES, D. J., *Neutron Transport Theory*, Pergamon, London, 1957.

HUGHES, D. J., *Neutron Optics*, Interscience, New York, 1954.

Elementary particles are the ultimate particles of matter, perhaps more correctly they are really the various forms which energy must take in order to become matter. Cosmic rays come from outer space with high speed, the primary ray being followed by secondary particles generated by the impact of the primary ray on air nuclei. They are the subject of study to see what agency is producing them, but also they can be used to determine the nature of interplanetary and geomagnetic fields. They have been the subject of a bibliography which, though done some years ago and not claiming to be complete, allows the reader to locate early work on the subject. The compilers were J. Timonno and J. A. Wheeler and the bibliography was published as "Guide to Literature of Elementary Particle Physics Including Cosmic Rays", *Amer. Sci.*, **37**, 202 (1949); **37**, 417 (1949). A more recent one is by B. Vitale, *A Bibliography on Heavy Mesons and Hypons*, North-Holland, Amsterdam, 1960.

Progress in the field is reviewed in the series *Progress in Elementary Particle and Cosmic Ray Physics*, North-Holland, Amsterdam, whilst for those seeking a short history Chen Ning Yang's *Elementary Particles*, Princeton University Press, Princeton, New Jersey, 1962, can be recommended.

Written for the layman is a text by D. T. Lewis, *Ultimate Particles of Matter*, Chantry, London, 1959. For the student, and based on a series of lectures, is Enrico Fermi's *Elementary Particles*, Oxford University Press, London, 1951. For the postgraduate student, A. Ramakrishnan has written *Elementary Particles and Cosmic Rays*, Pergamon, London, 1962. Another recommended text is by J. D. Jackson, *The Physics of Elementary Particles*, Oxford University Press, London, 1958.

Research work on cosmic rays is being undertaken at various universities and at the High Altitude Research Station, Jungfraujoch. This research station is controlled by the International Foundation of the High Altitude Research Station, Jungfraujoch. The Conseil Européenne pour la Recherche Nucléaire (CERN), Geneva, which was established as a co-operative research centre in 1954, also does research work with particular reference to phenomena involving very high energies leading to an understanding of the nature and properties of elementary particles. British research organizations working on elementary particle physics include the National Institute for Research in Nuclear Science.

Particle Accelerators [Dewey 539.73 Class]

A particle accelerator is a device for accelerating electrically charged particles to high energies.

Their primary use is as a research tool in nuclear and particle physics but they have also found use as producers of high volume rates of radiation for sterilization of food studies and for studies of radiation damage. The cost of particle accelerators puts them outside the reach of industrial laboratories and they are to be found only in the largest university, national and international laboratories such as the American University of California, Brookhaven National Laboratory and Stanford University, the British National Institute for Research in Nuclear Science and the international Conseil Européenne pour la Recherche Nucléaire (CERN).

The subject is described in the proceedings of various conferences organized by CERN, and in such recent texts as:

LIVINGOOD, J. M., *Principles of Cyclic Particle Accelerators*, Van Nostrand, Princeton, New Jersey, 1961.
LIVINGSTON, M. S. and BLEWATT, J. P., *Particle Accelerators*, McGraw-Hill, New York, 1962.

Mass Spectra [Dewey 539.744 Class]

A mass spectroscope is an instrument for determining the masses of atoms or molecules in a gas, liquid or solid. A beam of ions of the material being sent through electric or magnetic fields makes the ions of different masses travel along different paths and then impinge on a photographic plate or are detected electrically.

This method allows atomic masses to be measured very precisely and leads to it being considered as a method of manufacture of uranium 235 for the atomic bomb. Because chemical compounds have unique mass spectra, the mass spectroscope is widely employed as an analytical tool in the oil and other industries.

Several conferences on mass spectrometry have been held and most of these can be studied as the proceedings have been published:

Electromagnetically Enriched Isotopes and Mass Spectrometry: Proceedings of a Conference at Harwell, 13–16 September, 1955, Butterworth, London, 1956.

Mass Spectrometry: Report of a Conference by the Mass Spectrometry Panel of the Institute of Petroleum, Manchester, 20–21 April, 1950, Institute of Petroleum, London, 1952.

Mass Spectroscopy in Physics Research: Proceedings of the National Bureau of Standards Centennial Symposium, U.S. National Bureau of Standards, Washington D.C., 1953.

Recent advances in the field are given in the series *Advances in Mass Spectrometry*, Pergamon, London.

Books which have been well received include the classic by F. W. Aston, *Mass Spectra and Isotopes*, Arnold, London, 3rd ed., 1933, and more recent ones such as:

BARNARD, G. P., *Modern Mass Spectrometry*, Institute of Physics, London, 1953.

DUCKWORTH, H. E., *Mass Spectroscopy*, Cambridge University Press, London, 1958.

ROBERTSON, A. B. J., *Mass Spectrometry*, Wiley, New York, 1954.

Radioactivity and Radiation Research
[Dewey 539.752 Class]

Radioactivity is the result of the instability of the atomic nuclei of some atoms which, in order to reach a stable state, release energy in the form of radiation. The discovery of natural radioactivity by H. Becquerel in 1896 marked the birth of nuclear physics. The applications of radioactivity include the use of radioactive tracers to study chemical and physical processes, and in industry as control devices. Radiation research studies the effects of radiation on such materials as metals and biological materials.

Classical texts are those by Sir Ernest Rutherford, *Radiations from Radioactive Substances*, Cambridge University Press, London, 1930, and Sir James Chadwick, *Radioactivity and Radioactive Substances*, Pitman, London, 4th ed., 1953, whilst another recommended work is by J. M. Cork, *Radioactivity and Nuclear Physics*, Van Nostrand, Princeton, New Jersey, 3rd ed., 1957. UNESCO held an international conference on "Radioisotopes in Scientific Research" the proceedings of which have been published by Pergamon Press.

The effects of radiations on materials have been the subject of several bibliographies including *Effects of Neutron Irradiation in Non-Fissionable Metals and Alloys*, IAEA., Vienna, 1962, and textbooks such as:

BILLINGTON, D. S. and CRAWFORD, J. H., *Radiation Damage in Solids*, Princeton University Press, Princeton, New Jersey, 1961.

Periodicals

Periodicals covering radioactivity and radiation research include the monthly *International Journal of Applied Radiation and Isotopes* which covers isotope and radiation techniques, especially novel ones and those capable of wide application; the *British Journal of Radiology*, which is also published monthly and covers all aspects of the subject; and *Radiation Research*, which is the official organ of the Radiation Research Society and again is published monthly.

Societies and Research Organizations

Research is being undertaken by universities, hospitals, atomic energy research establishments and by other bodies such as the

National Bureau of Standards and the National Physical Laboratory. The actual work being undertaken can be found from the standard guides or annual reports.

America has many societies interested is this field, such as the Radiation Research Society which was first organized in 1952 to promote original research in the natural sciences relating to radiation and to promote the diffusion of knowledge. Annual meetings are held and the monthly periodical *Radiation Research* produced. Other American societies include the American College of Radiology, and the Radiological Society of North America. In Britain, there is the British Institute of Radiology which serves as a meeting place for radiologists, physicists, etc., and forms a centre for consultation and co-ordination of the medical, physical and biological aspects of radiology. Monthly meetings are held and the *British Journal of Radiology* published.

Considering the biological effects of radiation is the province of health physics. The health physicist ensures that workers using ionizing radiations do not receive doses above a certain permitted maximum. They establish what this maximum is, and ensure that the use and disposal of radioactive material does not lead to hazards. The hazards of radiation have been reviewed by the International Commission on Radiological Protection, and the Commission has published its recommendations of permissible doses of radiation. Another international commission of interest is the International Commission on Radiological Units and Measurement.

Societies active in health physics include the Health Physics Society which was organized in 1956 to improve the dissemination of information between individuals in the field and to improve public understanding of the problems and needs in radiation protection. The official journal of the Society is *Health Physics*, which is published monthly and has articles in any of four categories, research, engineering, applied and general.

Books of value include one in the Pergamon Progress in Nuclear Energy Series, *Series XII: Health Physics*, and the comprehensive handbook edited by H. Blatz, *Radiation Hygiene Handbook*,

McGraw-Hill, New York, 1959. The comprehensiveness of the
latter is well illustrated by its sections which cover exposure standards and radiation protection regulations; natural radioactive
background, ionizing radiation; sources of radiation; interaction
of radiation with matter; radiation attenuation data; laboratory
design; radiation detection and measurement; industrial applications; research applications; medical radiation applications;
determination of exposures; nuclear safety; radiation hygiene
chemistry; equipment for handling, storing and transport of radioactive materials; surface contamination and decontamination;
physiological effects of radiation; sampling equipment; liquid and
solid waste disposal; control of radioactive air pollution; personnel control; glossary of terms and reference data.

Plasma Physics [Dewey 539.764 Class]

A plasma is that part of a gas charge in which equal numbers
of electrons and positive ions are to be found, and thus is electrically neutral. The name was first introduced by Langmuir in
1929. Gas discharge plasmas offer a likely condition for high
enough temperatures to support controlled thermonuclear fusion.

Magnetohydrodynamics is concerned with the electromagnetic
phenomena in electrically conducting fluids, such as molten metals
and ionized gas. A thermonuclear reaction is the fusion of light
nuclei on energetic collision accompanied by the release of energy
and interest is centred on the development of a power producing
thermonuclear reactor.

Amongst the bibliographies available for consultation is one
compiled by J. D. Ramer and others, *Bibliography on Plasma
Physics and Magnetohydrodynamics and their Applications to
Controlled Thermonuclear Reactions*, University of Maryland,
College Park, Maryland, 1959. The arrangement of this bibliography, which covers the period 1937–59, is alphabetically by
author.

Magneto fluid-dynamics: Current Papers and Abstracts, Pergamon, London, 1962, was compiled by L. G. Napolitano and

G. Contursi and lists the references alphabetically by the name of the first author, then follows this by giving abstracts to some of the references. Author and subject indexes are included. Another recent bibliography is *Research on Controlled Thermonuclear Fusion*, International Atomic Energy Agency, Vienna, 1963. (IAEA Bibliographical Series No. 7.)

Symposia have been held covering this field, including the annual Lockheed symposium on magnetohydrodynamics, the fifth symposium being held in 1960. Its proceedings were published as *Radiations and Waves in Plasmas*, Stanford University Press, Stanford, California, 1961, under the editorship of M. Mitchner. Other symposia have been arranged by the International Atomic Energy Agency and other organizations and covered many aspects. The symposium held in Philadelphia in March 1961 dealt with flight applications, power conversion, diagnostics and fusion. Its proceedings, edited by C. Manual and N. W. Mather, are available as *Engineering Aspects of Magnetohydrodynamics*, Columbia University Press, New York, 1962. The papers which were presented at the 2nd United Nations Conference on the Peaceful Uses of Atomic Energy are available in the proceedings of the conference but have also been edited by A. P. Allis as *Nuclear Fusion*, Van Nostrand, Princeton, New Jersey, 1960.

Over the last few years there have been many books on plasma physics. The enquirer should have no trouble finding one to suit his needs. Two examples will illustrate the kind of material available. J. G. Linehard considers the behaviour of plasma in various fields of force, as studied by following the motion of single ions, in his book *Plasma Physics*, North-Holland, Amsterdam, 1960. A helpful bibliography is provided. A selection of previously unpublished articles covering theoretical and experimental investigations carried out at the Institute of Atomic Energy of the U.S.S.R. Academy of Sciences between 1951 and 1958 have been edited by M. A. Leontovich and published as *Plasma Physics and the Problem of Controlled Thermonuclear Reactions*, Pergamon, London, 1961.

The state of knowledge at a particular time is outlined in publications such as *Recent Research in Controlled Thermonuclear Fusion*, International Atomic Energy Agency, Vienna, and in the Progress in Nuclear Energy Series published by Pergamon, *Series XI: Plasma Physics and Thermonuclear Research*. More recent work can be found from the periodicals specializing in this field, such as *Journal of Nuclear Energy, Part C—Plasma Physics—Accelerators—Thermonuclear Research*, *Nuclear Fusion* and *Journal of Plasma Physics and Thermonuclear Fusion*.

Research in plasma physics is being undertaken in Britain by the United Kingdom Atomic Energy Authority at their Culham laboratories, and in America by the United States Atomic Energy Commission at Oak Ridge National Laboratory, by the Los Alamos Scientific Laboratory, the Princeton Plasma Physics Laboratory and the Lawrence Radiation Laboratory at Princeton, California. Magnetohydrodynamic studies are being pursued in Britain by such bodies as the Central Electricity Generating Board, United Kingdom Atomic Energy Authority and International Research and Development Ltd., and in other countries as well.

Particle Detection [Dewey 539.77 Class]

The instrumentation used in nuclear physics has been the subject of several fairly recent symposia, covering bubble chambers, photoelectric devices, semiconductor detectors, scintillation chambers and the like. Examples of the published proceedings are:

Nuclear Electronics: Proceedings of a Conference held in Belgrade, 15–20 May, 1961, International Atomic Energy Agency, Vienna, 1962.

Proceedings of an International Conference on Instrumentation for High Energy Physics, Berkeley, California, 12–24 September, 1960, Interscience, New York, 1961.

Proceedings of a Symposium on Nuclear Instruments, Harwell, 1961, Heywood, London, 1962.

Good examples of the kind of text available are the treatment by M. C. Nokes of Geiger-Müller counters, scalers, counting rate meters, etc., together with practical hints, in his *Radioactivity Measuring Instruments*, Heinemann, London, 1958, and the consideration of the physical properties and electronic techniques involved in cataloguing the electric signals from nuclear radiation detectors by R. L. Chase in his *Nuclear Pulse Spectroscopy*, McGraw-Hill, New York, 1961.

The specialized periodical for this field is *Nuclear Instruments and Methods*, which is issued monthly and covers the whole range of instruments and techniques in nuclear physics. Abstracts of articles from this periodical appear in *Chemical Abstracts*, *Physics Abstracts* (*Science Abstracts A*) and *Instrument Abstracts*.

Crystallography and X-Ray Diffraction
[Dewey 548, 548.83 Classes]

Crystallography is the science of the examination of crystalline material to obtain information on its structural units. X-Ray diffraction allows reflections to be obtained from any possible face of a crystal whether developed on the surface or not.

The newcomer to this subject will find historical treatments of value, and especially the story related by many of the scientists who were involved in the developments, *Fifty Years of X-ray Diffraction*, N.V.A. Oosthoek's Uitgeversmaatschappij, Utrecht, 1962.

Tables available for consultation, thus saving a possible long search, include:

KASPER, J. S. and LONSDALE, K. (eds.), *International Tables for X-ray Crystallography*, Kynoch Press, Birmingham, England, 1952, 1959. Volume 3 published in 1962 was edited by C. H. McGillavry, G. D. Rieck and K. Lonsdale.

WYCOFF, R. W. G., *Crystal Structures*, Interscience, New York, 1948–60. Issued in loose leaf form to facilitate additions and revisions this latter work aims to state the results of all available determinations of atomic positions in crystals. Bibliographical details are included.

Alphabetical and Grouped Numerical Index of X-ray Diffraction Data, American Society for Testing Materials, Philadelphia, 1950.

Tables for Conversion of X-ray Diffraction Angles to Interplanar Spacing, U.S. Government Printing Office, Washington D.C., 1950. (National Bureau of Standards Series 10.)

The classic texts are those by W. H. Bragg, who was the "father" of this science. His *X-rays and Crystal Structures*, Bell, London, 5th ed., 1925, is still required reading, as is the several-volume *The Crystalline State*, Bell, London, 1933–53, which has Sir W. H. Bragg and his son Sir W. L. Bragg as editors. The volumes comprising the series are:

Volume 1. BRAGG, SIR W. L., *A General Survey*, Bell, London, 1933.

Volume 2. JAMES, R. W., *The Optical Principles of the Diffraction of X-rays*, Bell, London, 1948.

Volume 3. LIPSON, H. and COCHRAN, W., *The Determination of Crystal Structures*, Bell, London, 1953.

Suitable books for those who have no background in the subject are:

BUERGER, M. J., *Elementary Crystallography*, Wiley, New York, 1956.

BUERGER, M. J., *Crystal Structure Analysis*, Wiley, New York, 1956.

BUNN, C. W., *Chemical Crystallography*, Oxford University Press, London, 2nd ed., 1961.

The research worker is catered for by such books as:

LIPSON, H. and TAYLOR, C. A., *Fourier Transforms and X-ray Diffraction*, Bell, London, 1958.

WOOSTER, W. A. *Diffuse X-ray Reflections from Crystals*, Oxford University Press, London, 1962.

Periodicals specific to the science include the Russian periodical *Kristallografiya*, which covers theory and experiment on crystal structure, lattice theory and diffraction studies. The American Institute of Physics issues a cover-to-cover translation of this periodical under the title *Soviet Physics—Crystallography*. The International Union of Crystallography publishes monthly *Acta Crystallographica*, which has material on all the branches of crystallography, solid-state physics and chemistry which have a structural basis. *Zeitschrift für Kristallographie, Kristallgeometrie, Kristallphysik, Kristallchemie* is the relevant German periodical.

Those societies interested in crystallography and X-ray diffraction, and from whom information can be obtained, include the International Union of Crystallography who work to promote international co-operation and publication of research in crystallography and also to facilitate standardization of methods and units. The Union is responsible for publishing *Acta Crystallographica, Structure Reports and International Tables for X-ray Crystallography*. The American Crystallographic Association was organized in 1949 to promote the study of the arrangement of atoms in matter, its causes, nature and consequences, and of the tools and methods used in such studies, and has as its British counterpart the section of the Institute of Physics and the Physical Society which is concerned with X-ray analysis.

Questions

1. Compare any two of the textbooks covering atomic physics.
2. What sort of information can be obtained from *Nuclear Data Sheets*?
3. What are the recent advances in mass spectrometry?

Index

HOW TO FIND OUT

A Guide to Sources of Information for all

G. CHANDLER. M.A., Ph.D., F.L.A., F.R.Hist.S.

City Librarian of Liverpool

The author has considerable experience in the recording, and dissemination, of information as City Librarian
of Liverpool, director of a technical information Centre
in association with the Department of Scientific and
Industrial Research, and extra-mural lecturer to the
Universities of Birmingham, Leeds and Liverpool.

In this book he gives a broad survey of information
sources on a wide variety of subjects, arranged by the
Dewey Classification. These are separate chapters on
guides to libraries, periodicals, reference books and
sources of information in general and special chapters
on the pure sciences, the earth and biological sciences,
technology, chemical technology, special industries
and the social sciences, literature, the arts, history,
geography and biography, etc. It is liberally illustrated
by 46 specimen pages from the most important sources
of information in all fields of knowledge.

HOW TO FIND OUT IN CHEMISTRY

A Guide to Sources of Information

C. R. BURMAN, B.A., F.L.A.

Technical Documentation Officer and
Librarian, Liverpool Public Libraries.

Contained in this book is an extremely useful summary of chemical information and sources of information of many types. For teachers, employment officers and personnel managers it provides a guide to the profession by outlining the careers available to qualified chemists, and explaining how such qualifications may be obtained. To the undergraduate and technical college student it gives a detailed exposition of how to make the best possible use of the international literature of chemistry, in both book and periodical form. It also outlines other sources of information that are available, such as the various international and professional societies and associations. The research worker will find the book useful as a résumé of the standard works in his own field of study, and of certain other fringe subjects. The guide to the use of libraries in general and to special chemical libraries in Great Britain and the United States will be of value to students of librarianship, and altogether the book will be an invaluable source of reference for all those mentioned above.